Rotating fields in general relativity

JAMAL NAZRUL ISLAM

Professor of Mathematics, University of Chittagong, Bangladesh

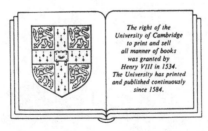

The right of the
University of Cambridge
to print and sell
all manner of books
was granted by
Henry VIII in 1534.
The University has printed
and published continuously
since 1584.

CAMBRIDGE UNIVERSITY PRESS

Cambridge

London New York New Rochelle

Melbourne Sydney

CAMBRIDGE UNIVERSITY PRESS
Cambridge, New York, Melbourne, Madrid, Cape Town, Singapore, São Paulo, Delhi

Cambridge University Press
The Edinburgh Building, Cambridge CB2 8RU, UK

Published in the United States of America by Cambridge University Press, New York

www.cambridge.org
Information on this title: www.cambridge.org/9780521113113

First published 1985
This digitally printed version 2009

A catalogue record for this publication is available from the British Library

Library of Congress Catalogue Card Number: 84–17071

ISBN 978-0-521-26082-4 hardback
ISBN 978-0-521-11311-3 paperback

To the memory of my parents
Mohammed Sirajul Islam and
Rahat Ara Begum

Contents

Contents

Preface

This book introduces the reader to research work on a particular aspect of rotating fields in general relativity. It should be accessible to someone with an elementary knowledge of general relativity, such as that obtained in an undergraduate course on general relativity at a British university. A person with some maturity in mathematical physics may be able to follow it without knowing general relativity, as I have given a brief introduction to the relevant aspects of general relativity in Chapter 1.

My intention has been to write a short book which can provide a relatively quick entry into some research topics. I have therefore made only a brief mention of some topics such as the important group theoretic generation of solutions by Kinnersley and others. A significant part of this book deals with interior solutions, for which these techniques are not yet applicable. I have also not touched upon Petrov classification of solutions as this is marginal to the problems considered in this book. The connecting link of the topics considered here is the Weyl–Lewis–Papapetrou form of the stationary axially symmetric metric, which is derived in detail in Chapters 1 and 2.

A significant part of the book is based on my own work and for this reason the book may be considered as too specialized. However, all research is specialized and I believe it is instructive for the beginning research worker to be shown a piece of work carried out to a certain stage of completion. Besides, I have tried, wherever possible, to bring out points of general interest.

In the earlier parts of the book I have usually carried out calculations explicitly. In the later parts I have left gaps which the reader is urged to fill him or herself. I frequently found myself running out of letters to use as symbols. I have sometimes used the same letters in different parts of the book, the different uses of which should be obvious from the context. I hope the reader will not mind this minor inconvenience.

Most of this book was written and much of the work on which parts of the book are based was done while I was at the City University, London. I am grateful to Prof. M.A. Jaswon and other members of the Mathematics Department there for their support. The book was completed during a visit to the Institute for Advanced Study, Princeton, New Jersey. I am grateful to Prof. F.J. Dyson and Prof. H. Woolf for hospitality there. I thank Alison Buttery, Catharine Rhubart, Ceinwen Sanderson and Veola Williams for typing parts of the manuscript. Lastly, I thank my wife Suraiya and my daughters Sadaf and Nargis for support and encouragement during the period in which this book was written.

Jamal Nazrul Islam

1

Introduction

1.1. Newtonian theory of gravitation

In Newton's theory of gravitation any sufficiently small piece of matter attracts any other sufficiently small piece of matter with a force which is inversely proportional to the square of the distance between the two pieces and which is proportional to the product of their masses. The constant of proportionality is G, Newton's gravitational constant. In this book we shall use units such that $G = 1$. From the inverse square law one can deduce that the gravitational field of a distribution of mass is described completely by a single function Φ of position, say of cartesian coordinates (x, y, z) and possibly of time t, which satisfies Poisson's equation inside matter, as follows:

$$\nabla^2\Phi \equiv \left(\frac{\partial^2}{\partial x^2} + \frac{\partial^2}{\partial y^2} + \frac{\partial^2}{\partial z^2}\right)\Phi(x, y, z, t) = -4\pi\varepsilon(x, y, z, t), \qquad (1.1)$$

where ε is the density of mass, which is also a function of x, y, z and possibly t. Outside matter, in empty space, Φ satisfies Laplace's equation, as follows:

$$\nabla^2\Phi = 0, \qquad (1.2)$$

a solution of which we will refer to as a harmonic function. The function Φ is referred to as the gravitational potential and has the physical significance that a particle of mass m placed in a gravitational field at the point (x, y, z) experiences a force $\mathbf{F}$ given by

$$\mathbf{F} = m\nabla\Phi. \qquad (1.3)$$

Equation (1.1) can be solved using the standard integral representation

$$\Phi(x, y, z, t) = \int \frac{\varepsilon(x', y', z', t)\, dx'\, dy'\, dz'}{[(x - x')^2 + (y - y')^2 + (z - z')^2]^{1/2}}, \qquad (1.4)$$

where the integration is over the region in which ε is non-zero. However, (1.4) is usually difficult to evaluate and other simpler methods are used to

Introduction

arrive at the potential Φ, especially in situations involving symmetries.

In this book we shall be concerned with situations of axial or cylindrical symmetry. These symmetries have to be defined rigorously in general relativity, but it is useful to have in mind the simpler definitions of these symmetries in the Newtonian situation. Even here we shall take a somewhat intuitive approach lacking in rigour but it will be simple and adequate for our purpose. We first define cylindrical polar coordinates. A point P has cylindrical polar coordinates (ρ, ϕ, z) related to its cartesian coordinates (x, y, z) as follows:

$$x = \rho \cos \phi, \quad y = \rho \sin \phi. \tag{1.5}$$

Thus ρ is the distance of the point P from the z-axis and ϕ is the angle which

Fig. 1.1. Illustration of cylindrical polar coordinates.

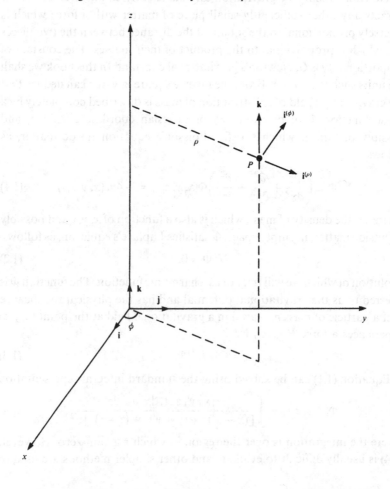

the plane containing the z-axis and the point P makes with the plane $y = 0$ (Fig. 1.1). The angle ϕ is often referred to as the azimuthal angle. A scalar function of position $f(x, y, z)$, is said to be axially symmetric (with the z-axis as the axis of symmetry) if, when expressed in terms of coordinates (ρ, ϕ, z), it is independent of ϕ, that is,

$$f(\rho\cos\phi, \rho\sin\phi, z) = F(\rho, z), \tag{1.6}$$

where $F(\rho, z)$ is a function of ρ and z only. Thus axially symmetric functions have rotational symmetry about the z-axis. There are different ways of interpreting this last statement. Consider a circle passing through P with its centre on the z-axis and its plane parallel to the plane $z = 0$. An axially symmetric function has the same value at all points of this circle, since this circle is described by fixed values of ρ and z. Thus the surfaces $F(\rho, z) = $ constant have the property that they look the same in position and shape if they are rotated by any given angle about the z-axis, that is, they are invariant under a rotation of the coordinate system about the z-axis.

Consider now a three-dimensional vector field $\mathbf{H}(x, y, z)$ with components $H^{(x)}$, $H^{(y)}$, $H^{(z)}$ along the x, y and z-axis respectively, that is,

$$\mathbf{H}(x, y, z) = \mathbf{i}H^{(x)}(x, y, z) + \mathbf{j}H^{(y)}(x, y, z) + \mathbf{k}H^{(z)}(x, y, z), \tag{1.7}$$

where $\mathbf{i}, \mathbf{j}, \mathbf{k}$ are unit vectors along the x, y, z axes respectively. The vector field $\mathbf{H}$ is axially symmetric if its components when expressed in terms of the triad of unit vectors $(\mathbf{i}^{(\rho)}, \mathbf{i}^{(\phi)}, \mathbf{k})$ are independent of ϕ. The unit vectors $\mathbf{i}^{(\rho)}$ and $\mathbf{i}^{(\phi)}$ are defined as follows:

$$\mathbf{i}^{(\rho)} = \mathbf{i}\cos\phi + \mathbf{j}\sin\phi, \quad \mathbf{i}^{(\phi)} = -\mathbf{i}\sin\phi + \mathbf{j}\cos\phi. \tag{1.8}$$

The vector $\mathbf{i}^{(\rho)}$ points radially away from the z-axis and is parallel to the plane $z = 0$, while $\mathbf{i}^{(\phi)}$ is also parallel to this plane and is perpendicular to $\mathbf{i}^{(\rho)}$, pointing in the direction of increasing ϕ (Fig. 1.1). Thus $\mathbf{H}$ is axially symmetric if

$$\mathbf{H} = \mathbf{i}^{(\rho)}H^{(\rho)}(\rho, z) + \mathbf{i}^{(\phi)}H^{(\phi)}(\rho, z) + \mathbf{k}H^{(z)}(\rho, z). \tag{1.9}$$

Note that if $\mathbf{H}$ is axially symmetric (so that $H^{(\rho)}$, $H^{(\phi)}$, $H^{(z)}$ are independent of ϕ) its components in terms of $\mathbf{i}, \mathbf{j}, \mathbf{k}$ do depend on ϕ in the following simple manner:

$$\mathbf{H} = \mathbf{i}(\cos\phi H^{(\rho)} - \sin\phi H^{(\phi)}) + \mathbf{j}(\sin\phi H^{(\rho)} + \cos\phi H^{(\phi)}) + \mathbf{k}H^{(z)}. \tag{1.10}$$

From (1.9) it is clear that an axially symmetric vector field is in a suitable sense invariant under rotation of the axes about the z-axis by any given angle.

Cylindrical symmetry can be defined in the same manner as above if we make the various functions independent of z in addition to being independent of ϕ. Thus a scalar function $f(x, y)$ (which is independent of z) is cylindrically symmetric (with the z-axis as the axis of symmetry) if, when expressed in terms of coordinates ρ, ϕ, z it is a function of ρ only:

$$f(\rho \cos \phi, \rho \sin \phi) = F(\rho). \qquad (1.11)$$

Thus if a and b are constants, the function $a\rho^2$ is cylindrically symmetric but the function $a\rho^2 + bz^2$, although axially symmetric, is not cylindrically symmetric. The vector field $\mathbf{H}$ is cylindrically symmetric if the components $H^{(\rho)}$, $H^{(\phi)}$ and $H^{(z)}$ are functions of ρ only. Cylindrically symmetric systems are invariant not only under a rotation about the z-axis, but they are also invariant under a translation parallel to the z-axis. These remarks may seem obvious but they help to fix ideas for the more complicated situations encountered later in the book and besides, confusion does arise sometimes between axial and cylindrical symmetry.

As the title of the book implies, we shall be concerned with rotating systems. It is therefore pertinent to consider a simple rotating system in Newtonian theory, namely, the case of uniformly (rigidly) rotating inviscid homogeneous fluid, where the rotation is steady, that is, independent of time. As is well known, the boundary of such a fluid mass is an oblate spheroid, which is an ellipsoid with two equal axes, these equal axes being greater than the third axis. A prolate spheroid is one in which the equal axes are smaller than the third axis. A typical portion of the material of the fluid mass is kept in equilibrium by gravitational, pressure and centrifugal forces. We need not concern ourselves with the equations governing these forces (see, for example, Chandrasekhar, 1969). If we assume the centre of the spheroid to be at the centre of the coordinate system, the gravitational potential inside the matter is given by

$$\Phi(\rho, z) = a\rho^2 + bz^2 + \Phi_0, \qquad (1.12)$$

where a, b, Φ_0 are constants, and Φ is given by a more complicated but explicitly known function of ρ and z outside the matter. From (1.1) we see that a, b and ε (which is constant in this case) are related by

$$\nabla^2\Phi \equiv \Phi_{\rho\rho} + \rho^{-1}\Phi_\rho + \Phi_{zz} = 4a + 2b = -4\pi\varepsilon, \qquad (1.13)$$

where $\Phi_\rho \equiv \partial\Phi/\partial\rho$, etc. The interior and exterior potentials join smoothly at the boundary of the matter distribution, where the pressure is zero. By joining smoothly we mean that the potential and its partial derivatives with respect to ρ and z are continuous at the boundary.

We will now describe one of the fundamental differences between a Newtonian rotating system and a general relativistic one. Let a test particle of mass m be released from rest at a great distance from the rotating mass considered above in the equatorial plane $z = 0$ (Fig. 1.2). According to (1.3) the force on the test particle is given by

$$\mathbf{F} = m\nabla\Phi = \left\{ m\left(\mathbf{i}^{(\rho)}\frac{\partial}{\partial\rho} + \mathbf{i}^{(\phi)}\rho^{-1}\frac{\partial}{\partial\phi} + \mathbf{k}\frac{\partial}{\partial z} \right)\Phi \right\}_{z=0} \tag{1.14}$$

Since the system is axially symmetric about the z-axis, Φ is independent of ϕ so that the coefficient of $\mathbf{i}^{(\phi)}$ in (1.14) vanishes. This is true for all z and not just for $z = 0$. Thus for any position of the test particle there is no transverse force on the particle in the $\mathbf{i}^{(\phi)}$ direction. Returning to (1.14), the system has reflection symmetry about the plane $z = 0$, so Φ depends on z through z^2 only (recall that (1.12) is just the interior potential; the exterior potential is given by a different function), so that

$$\left\{ \frac{\partial\Phi}{\partial z} \right\}_{z=0} = \left\{ 2z\frac{\partial\Phi}{\partial z^2} \right\}_{z=0} = 0. \tag{1.15}$$

Thus

$$\mathbf{F} = m\frac{\partial\Phi}{\partial\rho}(\rho, 0)\mathbf{i}^{(\rho)}, \tag{1.16}$$

so that the force on the test particle is radial and it will follow the dashed straight line through the centre of the mass (Fig. 1.2). Similar considerations also apply for any rotating body which has axial symmetry and reflection symmetry on the plane $z = 0$, but we consider the above example for definiteness. Considering now the corresponding situation in general relativity, that is, the case of a steadily and rigidly rotating homogeneous

Fig. 1.2. Path of a test particle released from rest in the equatorial plane of a rotating mass in Newtonian gravitation (dashed line) and in general relativity (continuous line).

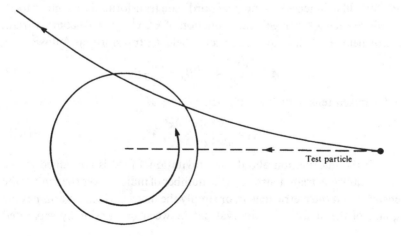

Test particle

inviscid fluid, the first thing to note is that the surface of the mass is no longer a spheroid. In fact due to the non-Euclidean nature of the geometry it is difficult to characterize the surface in coordinates, but it is still true that the pressure vanishes on this surface. The concept of the equatorial plane and the fact that the system has reflection symmetry about this plane can be taken over to general relativity and one can ask what will happen to a test particle if it is released from rest on the equatorial plane. In this case there will be a transverse force on the test particle because of the rotation of the central mass and the particle will follow the continuous line (Fig. 1.2). This phenomenon is related to what is referred to as 'inertial dragging' and will be discussed later. Thus in general relativity matter in motion exerts a force akin to magnetic forces exerted by electric charges in motion. This is not true in Newtonian gravitation.

1.2. Summary of general relativity

The reader is assumed to be familiar with the elements of general relativity but we shall give here a brief review as a reminder of the main results and to collect together formulae some of which will be useful later in the book.

General relativity is formulated in a four-dimensional Riemannian space in which points are labelled by a general non-inertial coordinate system (x^0, x^1, x^2, x^3), often written as x^μ ($\mu = 0, 1, 2, 3$) (we use the covention that Greek indices take values 0, 1, 2, 3 and repeated Greek indices are to be summed over these values unless otherwise stated; the meaning of other indices will be specified as they arise). Several coordinate patches may be necessary to cover the whole of space–time. The space has three spatial and one time-like dimension. Under a coordinate transformation from x^μ to x'^μ (in which each x'^μ is in general a function of x^0, x^1, x^2, x^3) a contravariant vector field A^μ and a covariant vector field B_μ transform as follows:

$$A'^\mu = \frac{\partial x'^\mu}{\partial x^\nu} A^\nu, \quad B'_\mu = \frac{\partial x^\nu}{\partial x'^\mu} B_\nu,$$ (1.17)

and a mixed tensor such as $A^\mu_{\ \nu\lambda}$ transforms as

$$A'^\mu_{\ \nu\lambda} = \frac{\partial x'^\mu}{\partial x^\rho} \frac{\partial x^\sigma}{\partial x'^\nu} \frac{\partial x^\tau}{\partial x'^\lambda} A^\rho_{\ \sigma\tau},$$ (1.18)

etc. All the information about the gravitational field is contained in the second rank covariant tensor $g_{\mu\nu}$ (the number of indices gives the rank of the tensor) called the metric tensor, or simply the metric, which determines the square of the space–time interval ds^2 between infinitesimally separated events or points x^μ and $x^\mu + dx^\mu$ as follows ($g_{\mu\nu} = g_{\nu\mu}$):

$$ds^2 = g_{\mu\nu}dx^\mu dx^\nu. \tag{1.19}$$

The contravariant tensor corresponding to $g_{\mu\nu}$ is denoted by $g^{\mu\nu}$ and is defined by

$$g_{\mu\nu}g^{\nu\lambda} = \delta_\mu^\lambda \tag{1.20}$$

where δ_μ^λ is the Kronecker delta, which equals unity if $\lambda = \mu$ (no summation) and zero otherwise. Indices can be raised or lowered by using the metric tensor as follows:

$$A^\mu = g^{\mu\nu}A_\nu, \quad A_\mu = g_{\mu\nu}A^\nu. \tag{1.21}$$

The generalization of ordinary (partial) differentiation to Riemannian space is given by covariant differentiation denoted by a semicolon and defined for a contravariant and a covariant vector as follows:

$$A^\mu_{\;;\nu} = \frac{\partial A^\mu}{\partial x^\nu} + \Gamma^\mu_{\nu\lambda}A^\lambda, \tag{1.22a}$$

$$A_{\mu;\nu} = \frac{\partial A_\mu}{\partial x^\nu} - \Gamma^\lambda_{\mu\nu}A_\lambda. \tag{1.22b}$$

Here the $\Gamma^\mu_{\nu\lambda}$ are called Christoffel symbols; they have the property $\Gamma^\mu_{\nu\lambda} = \Gamma^\mu_{\lambda\nu}$ and are given in terms of the metric tensor as follows:

$$\Gamma^\mu_{\nu\lambda} = \tfrac{1}{2}g^{\mu\sigma}(g_{\sigma\nu,\lambda} + g_{\sigma\lambda,\nu} - g_{\nu\lambda,\sigma}), \tag{1.23}$$

where a comma denotes partial differentiation with respect to the corresponding variable: $g_{\sigma\nu,\lambda} \equiv \partial g_{\sigma\nu}/\partial x^\lambda$. For covariant differentiation of tensors of higher rank, there is a term corresponding to each contravariant index analogous to the second terms in (1.22a) and a term corresponding to each covariant index analogous to the second term in (1.22b) (with a negative sign). Equation (1.23) has the consequence that the covariant derivative of the metric tensor vanishes:

$$g_{\mu\nu;\lambda} = 0, \quad g^{\mu\nu}_{\;\;;\lambda} = 0. \tag{1.24}$$

Under a coordinate transformation from x^μ to x'^μ the $\Gamma^\mu_{\nu\lambda}$ transform as follows:

$$\Gamma'^\mu_{\nu\lambda} = \frac{\partial x'^\mu}{\partial x^\rho}\frac{\partial x^\sigma}{\partial x'^\nu}\frac{\partial x^\tau}{\partial x'^\lambda}\Gamma^\rho_{\sigma\tau} + \frac{\partial^2 x^\sigma}{\partial x'^\nu \partial x'^\lambda}\frac{\partial x'^\mu}{\partial x^\sigma}, \tag{1.25}$$

so that the $\Gamma^\mu_{\nu\lambda}$ do not form components of a tensor since the transformation law (1.25) is different to that of a tensor (see (1.18)).

For any covariant vector A_μ it can be shown that

$$A_{\mu;\nu;\lambda} - A_{\mu;\lambda;\nu} = A_\sigma R^\sigma_{\;\mu\nu\lambda}, \tag{1.26}$$

where $R^\sigma{}_{\mu\nu\lambda}$ is the Riemann tensor defined by

$$R^\sigma{}_{\mu\nu\lambda} = \Gamma^\sigma{}_{\mu\lambda,\nu} - \Gamma^\sigma{}_{\mu\nu,\lambda} + \Gamma^\sigma{}_{\alpha\nu}\Gamma^\alpha{}_{\mu\lambda} - \Gamma^\sigma{}_{\alpha\lambda}\Gamma^\alpha{}_{\mu\nu}. \tag{1.27}$$

The Riemann tensor has the following symmetry properties:

$$R_{\sigma\mu\nu\lambda} = -R_{\mu\sigma\nu\lambda} = -R_{\sigma\mu\lambda\nu}, \tag{1.28a}$$

$$R_{\sigma\mu\nu\lambda} = R_{\nu\lambda\sigma\mu}, \tag{1.28b}$$

$$R_{\sigma\mu\nu\lambda} + R_{\sigma\lambda\mu\nu} + R_{\sigma\nu\lambda\mu} = 0, \tag{1.28c}$$

and satisfies the Bianchi identity:

$$R^\sigma{}_{\mu\nu\lambda;\rho} + R^\sigma{}_{\mu\rho\nu;\lambda} + R^\sigma{}_{\mu\lambda\rho;\nu} = 0. \tag{1.29}$$

The Ricci tensor $R_{\mu\nu}$ is defined by

$$R_{\mu\nu} = g^{\lambda\sigma}R_{\lambda\mu\sigma\nu} = R^\sigma{}_{\mu\sigma\nu}. \tag{1.30}$$

From (1.27) and (1.30) it follows that $R_{\mu\nu}$ is given by

$$R_{\mu\nu} = \Gamma^\lambda{}_{\mu\nu,\lambda} - \Gamma^\lambda{}_{\mu\lambda,\nu} + \Gamma^\lambda{}_{\mu\nu}\Gamma^\sigma{}_{\lambda\sigma} - \Gamma^\sigma{}_{\mu\lambda}\Gamma^\lambda{}_{\nu\sigma}. \tag{1.31}$$

Let the determinant of $g_{\mu\nu}$ considered as a matrix be denoted by g. Then another expression for $R_{\mu\nu}$ is given as follows:

$$R_{\mu\nu} = \frac{1}{(-g)^{1/2}}\,[\Gamma^\lambda{}_{\mu\nu}(-g)^{1/2}]_{,\lambda} - [\log(-g)^{1/2}]_{,\mu\nu} - \Gamma^\sigma{}_{\mu\lambda}\Gamma^\lambda{}_{\nu\sigma}. \tag{1.31a}$$

This follows from the fact that from (1.23) and the properties of matrices one can show that

$$\Gamma^\lambda{}_{\mu\lambda} = [\log(-g)^{1/2}]_{,\mu}. \tag{1.32}$$

From (1.31a) it follows that $R_{\mu\nu} = R_{\nu\mu}$. There is no agreed convention for the sign of the Riemann and Ricci tensors – some authors define these with opposite sign to (1.27) and (1.31). The Ricci scalar R is defined by

$$R = g^{\mu\nu}R_{\mu\nu}. \tag{1.33}$$

By contracting the Bianchi identity on the pair of indices $\mu\nu$ and $\sigma\rho$ (that is, multiplying it by $g^{\mu\nu}$ and $g^{\sigma\rho}$) one can deduce the identity

$$(R^{\mu\nu} - \tfrac{1}{2}g^{\mu\nu}R)_{;\nu} = 0. \tag{1.34}$$

The tensor $G^{\mu\nu} \equiv (R^{\mu\nu} - \tfrac{1}{2}g^{\mu\nu}R)$ is sometimes called the Einstein tensor.

   We are now in a position to write down the fundamental equations of general relativity. These are Einstein's equations given by

$$R_{\mu\nu} - \tfrac{1}{2}g_{\mu\nu}R = 8\pi T_{\mu\nu}, \tag{1.35}$$

where $T_{\mu\nu}$ is the energy-momentum tensor of the source producing the gravitational field. In (1.35) and throughout the following we use units such that Newton's gravitational constant and the velocity of light are both equal to unity. For a perfect fluid $T_{\mu\nu}$ takes the following form:

$$T^{\mu\nu} = (\varepsilon + p)u^\mu u^\nu - p g^{\mu\nu}, \tag{1.36}$$

where ε is the mass-energy density (note the contrast to (1.1) where ε is just the mass density), p is the pressure and u^μ is the four-velocity of matter given by

$$u^\mu = \frac{\mathrm{d}x^\mu}{\mathrm{d}s}, \tag{1.37}$$

where $x^\mu(s)$ describes the world-line of the matter in terms of the proper time s along the world-line. The energy-momentum tensor for the electromagnetic field will be considered in a later chapter. From (1.34) we see that Einstein's equations (1.35) are compatible with the following equation

$$T^{\mu\nu}_{\ ;\nu} = 0, \tag{1.38}$$

which is the equation for the conservation of mass-energy and momentum.

The equations of motion of a particle in a gravitational field are given by the geodesic equations:

$$\frac{\mathrm{d}^2 x^\mu}{\mathrm{d}s^2} + \Gamma^\mu_{\nu\lambda} \frac{\mathrm{d}x^\nu}{\mathrm{d}s} \frac{\mathrm{d}x^\lambda}{\mathrm{d}s} = 0. \tag{1.39}$$

Geodesics can also be introduced through the concept of parallel transfer. We will consider this concept and related matters in the next section.

This completes our brief survey of some of the elements of general relativity that we shall assume knowledge of in the following. In the rest of this chapter we shall cover some additional topics which are pertinent to the following chapters.

1.3. Curves in Riemannian space

A curve in Riemannian space is defined by points $x^\mu(\lambda)$ where x^μ are suitably differentiable functions of the real parameter λ, varying over some interval of the real line. Consider a coordinate transformation from x^μ to x'^μ. The set of quantities $\mathrm{d}x^\mu/\mathrm{d}\lambda$ transform to $\mathrm{d}x'^\mu/\mathrm{d}\lambda$ given by

$$\frac{\mathrm{d}x'^\mu}{\mathrm{d}\lambda} = \frac{\partial x'^\mu}{\partial x^\nu} \frac{\mathrm{d}x^\nu}{\mathrm{d}\lambda}, \tag{1.40}$$

that is, $\mathrm{d}x^\mu/\mathrm{d}\lambda$ are components of a contravariant vector. This is called the tangent vector to the curve. The curve, or a portion of it, is time-like, light-like or space-like according as to whether $g_{\mu\nu}\,\mathrm{d}x^\mu/\mathrm{d}\lambda\,\mathrm{d}x^\nu/\mathrm{d}\lambda > 0, = 0,$ or < 0.

(At any point $g_{\mu\nu}$ can be reduced to the diagonal form $(1, -1, -1, -1)$ by a suitable transformation.) The length of the time-like or space-like curve from $\lambda = \lambda_1$ to $\lambda = \lambda_2$ is given by:

$$L_{12} = \int_{\lambda_1}^{\lambda_2} \left(\left| g_{\mu\nu} \frac{dx^\mu}{d\lambda} \frac{dx^\nu}{d\lambda} \right| \right)^{1/2} d\lambda. \qquad (1.41)$$

The length of a light-like curve is zero. For an arbitrary vector field Y^μ its covariant derivative along the curve is the vector (defined along the curve) $Y^\mu{}_{;\nu}(dx^\nu/d\lambda)$. A similar definition can be given for the covariant derivative of an arbitrary tensor field along the curve. The vector field Y^μ is said to be parallelly transported along the curve if

$$Y^\mu{}_{;\nu} \frac{dx^\nu}{d\lambda} = Y^\mu{}_{,\nu} \frac{dx^\nu}{d\lambda} + \Gamma^\mu_{\nu\sigma} Y^\sigma \frac{dx^\nu}{d\lambda}$$

$$= \frac{dY^\mu}{d\lambda} + \Gamma^\mu_{\nu\sigma} Y^\sigma \frac{dx^\nu}{d\lambda} = 0. \qquad (1.42)$$

A similar definition holds for tensors. Given any curve $x^\mu(\lambda)$ with end points $\lambda = \lambda_1$ and $\lambda = \lambda_2$ the theory of solutions of ordinary differential equations shows that if the $\Gamma^\mu_{\nu\sigma}$ are suitably differentiable functions of the x^μ one obtains a unique tensor at $\lambda = \lambda_2$ by parallelly transferring any given tensor from $\lambda = \lambda_1$ along $x^\mu(\lambda)$, if the latter are differentiable in λ. A particular case is the covariant derivative of the tangent vector itself along $x^\mu(\lambda)$. The curve is said to be a geodesic curve if the tangent vector is transported parallelly along the curve, that is (putting $Y^\mu = dx^\mu/d\lambda$ in (1.42)) if

$$\frac{d^2x^\mu}{d\lambda^2} + \Gamma^\mu_{\nu\sigma} \frac{dx^\nu}{d\lambda} \frac{dx^\sigma}{d\lambda} = 0. \qquad (1.43)$$

A geodesic, or a portion of it, can be time-like, light-like or space-like according to the type of curve it is.

Two vector fields V^μ, W^μ are normal or orthogonal to each other if $g_{\mu\nu}V^\mu W^\nu = 0$. If V^μ is time-like and orthogonal to W^μ then the latter is necessarily space-like. A space-like three-surface is a surface defined by $f(x^0, x^1, x^2, x^3) = 0$ such that $g^{\mu\nu}f_{,\mu}f_{,\nu} > 0$ when $f = 0$. The unit normal vector to this surface is given by $n^\mu = (g^{\alpha\beta}f_{,\alpha}f_{,\beta})^{-1/2} g^{\mu\nu}f_{,\nu}$.

Given a vector field ξ^μ, one can define a set of curves filling all space such that the tangent vector to any curve of this set at any point coincides with the value of the vector field at that point. This is done by solving the set of first order differential equations.

$$\frac{dx^\mu}{d\lambda} = \xi^\mu(x(\lambda)), \qquad (1.44)$$

where on the right hand side we have put x for all four components of the coordinates. This set of curves is referred to as the congruence of curves generated by the given vector field. In general there is a unique member of this congruence passing through any given point. A particular member of the congruence is sometimes referred to as an orbit. Consider now the vector field given by $(\xi^0, \xi^1, \xi^2, \xi^3) = (1, 0, 0, 0)$. From (1.44) we see that the congruence of this vector field is the set of curves given by

$$(x^0 = \lambda, x^1 = \text{constant}, x^2 = \text{constant}, x^3 = \text{constant}). \qquad (1.45)$$

This vector field is also referred to as the vector field $\partial/\partial x^0$. One similarly defines the vector fields $\partial/\partial x^1$, $\partial/\partial x^2$, $\partial/\partial x^3$. That is, corresponding to the coordinate system x^μ we have the four contravariant vector fields $\partial/\partial x^\mu$. A general vector field X^μ can be written without components in terms of $\partial/\partial x^\mu$ as follows:

$$\mathbf{X} = X^\mu \frac{\partial}{\partial x^\mu}. \qquad (1.46)$$

This is related to the fact that contravariant vectors at any point can be regarded as operators acting on differentiable functions $f(x^0, x^1, x^2, x^3)$; when the vector acts on the function, the result is the derivative of the function in the direction of the vector field, as follows:

$$\mathbf{X}(f) = X^\mu \frac{\partial f}{\partial x^\mu}. \qquad (1.47)$$

As is well known, differential geometry and, correspondingly, general relativity can be developed independently of coordinates and components. We shall not be concerned with this approach except incidentally. Whenever we use this approach we shall specify the relevant points as we have done in (1.47).

1.4. Killing vectors

Einstein's exterior equations $R_{\mu\nu} = 0$ (obtained from (1.35) by setting $T_{\mu\nu} = 0$) are a set of coupled non-linear partial differential equations for the ten unknown functions $g_{\mu\nu}$. The interior equations (1.35) may involve other unknown functions such as the mass-energy density and the pressure. Because of the freedom to carry out general coordinate transformations one can in general impose four conditions on the ten functions $g_{\mu\nu}$. Later we will show explicitly how this is done in a case involving symmetries. In most situations of physical interest one has space–time symmetries which reduce further the number of unknown functions. To determine the simplest form of the metric (that is, the form of $g_{\mu\nu}$) when one has a given space–time symmetry is a non-trivial problem. For example in Newtonian theory spherical symmetry is usually defined by a centre and the property that all points at any given distance from the centre are equivalent. This definition cannot be taken over directly to general relativity. In the latter, 'distance' is defined by the metric to begin with and, for example, the 'centre' may not be accessible to physical measurement, as is indeed the case in the Schwarzschild geometry (see Section 2.4). One therefore has to find some coordinate independent and covariant manner of defining space–time symmetries such as axial symmetry and stationarity. This is done with the help of Killing vectors, which we will now consider. In some cases there is a less rigorous but simpler way of deriving the metric which we will also consider.

In the following we will sometimes write x, y, x' for x^μ, y^μ, x'^μ respectively. A metric $g_{\mu\nu}(x)$ is form-invariant under a transformation from x^μ to x'^μ if $g'_{\mu\nu}(x')$ is the same function of x'^μ as $g_{\mu\nu}(x)$ is of x^μ. For example, the Minkowski metric is form-invariant under a Lorentz transformation. Thus

$$g'_{\mu\nu}(y) = g_{\mu\nu}(y), \quad \text{all } y. \tag{1.48}$$

Therefore

$$g_{\mu\nu}(x) = \frac{\partial x'^\rho}{\partial x^\mu}\frac{\partial x'^\sigma}{\partial x^\nu} g'_{\rho\sigma}(x') = \frac{\partial x'^\rho}{\partial x^\mu}\frac{\partial x'^\sigma}{\partial x^\nu} g_{\rho\sigma}(x'). \tag{1.49}$$

The transformation from x^μ to x'^μ in this case is called an isometry of $g_{\mu\nu}$. Consider an infinitesimal isometry transformation from x^μ to x'^μ defined by

$$x'^\mu = x^\mu + \alpha \xi^\mu(x), \tag{1.50}$$

with α constant and $|\alpha| \ll 1$. Substituting in (1.49) and neglecting terms involving α^2 we arrive at the following equation (see e.g. Weinberg (1972)):

$$g_{\mu\sigma}\frac{\partial \xi^\mu}{\partial x^\rho} + g_{\rho\mu}\frac{\partial \xi^\mu}{\partial x^\sigma} + \frac{\partial g_{\rho\sigma}}{\partial x^\mu}\xi^\mu = 0. \tag{1.51}$$

With the use of (1.22b) and (1.23), the equation (1.51) can be written as follows:

$$\xi_{\sigma;\rho} + \xi_{\rho;\sigma} = 0. \tag{1.52}$$

Equation (1.52) is Killing's equation and a vector field ξ^μ satisfying it is called a Killing vector of the metric $g_{\mu\nu}$. Thus if there exists a solution of (1.52) for a given $g_{\mu\nu}$, then the corresponding ξ^μ represents an infinitesimal isometry of the metric $g_{\mu\nu}$ and implies that the metric has a certain symmetry. Since (1.52) is covariantly expressed, that is, it is a tensor equation, if the metric has an isometry in a given coordinate system, in any transformed coordinate system the transformed metric will also have a corresponding isometry. This is important because often a metric can look quite different in different coordinate systems.

To give an example of a Killing vector, we consider a situation in which the metric is independent of one of the four coordinates. To fix ideas, we choose this coordinate to be x^0, which we take to be time-like, that is, the lines $(x^0 = \lambda, x^1 = \text{constant}, x^2 = \text{constant}, x^3 = \text{constant})$ for varying λ are time-like lines. In general, $g_{\mu\nu}$ being independent of x^0 means that the gravitational field is stationary, that is, it is produced by sources whose state of motion does not change with time. In this case we have

$$\frac{\partial g_{\mu\nu}}{\partial x^0} \equiv g_{\mu\nu,0} = 0. \tag{1.53}$$

Consider now the vector field ξ^μ given by

$$(\xi^0, \xi^1, \xi^2, \xi^3) = (1, 0, 0, 0), \tag{1.54}$$

with $\xi_\mu = g_{\mu\nu}\xi^\nu = g_{\mu 0}$. We have

$$\begin{aligned}
\xi_{\mu;\nu} + \xi_{\nu;\mu} &= \xi_{\mu,\nu} + \xi_{\nu,\mu} - g^{\lambda\sigma}(g_{\sigma\mu,\nu} + g_{\sigma\nu,\mu} - g_{\mu\nu,\sigma})\xi_\lambda \\
&= g_{\mu 0,\nu} + g_{\nu 0,\mu} - \xi^\sigma(g_{\sigma\mu,\nu} + g_{\sigma\nu,\mu} - g_{\mu\nu,\sigma}) \\
&= g_{\mu\nu,0} = 0, \tag{1.55}
\end{aligned}$$

using (1.53) and (1.54). Thus if (1.53) is satisfied, the vector (1.54) gives a solution to Killing's equation. In other words, if the metric admits the Killing vector (1.54), then (1.53) is satisfied and the metric is stationary. A similar result can be established for any of the other three coordinates.

We now derive a property of Killing vectors which we will use later. Let $\xi^{(1)\mu}$ and $\xi^{(2)\mu}$ be two linearly independent solutions of Killing's equation (1.52). We define the commutator of these two Killing vectors as the vector ζ^μ given by

$$\zeta^\mu = \xi^{(1)\mu}{}_{;\lambda}\xi^{(2)\lambda} - \xi^{(2)\mu}{}_{;\lambda}\xi^{(1)\lambda}. \tag{1.56}$$

In coordinate independent notation the commutator of $\xi^{(1)}$ and $\xi^{(2)}$ is written as $[\xi^{(1)}, \xi^{(2)}]$. In fact because of the symmetry of the Christoffel symbols the covariant derivatives in (1.56) can be replaced by ordinary derivatives. We will now show that ζ^μ is also a Killing vector, that is,

$$\zeta_{\mu;\nu} + \zeta_{\nu;\mu} = 0. \tag{1.57}$$

Now

$$\begin{aligned}
\zeta_{\mu;\nu} + \zeta_{\nu;\mu} &= \xi^{(1)}_{\mu;\lambda;\nu}\xi^{(2)\lambda} + \xi^{(1)}_{\mu;\lambda}\xi^{(2)\lambda}{}_{;\nu} - \xi^{(2)}_{\mu;\lambda;\nu}\xi^{(1)\lambda} \\
&\quad - \xi^{(2)}_{\mu;\lambda}\xi^{(1)\lambda}{}_{;\nu} + \xi^{(1)}_{\nu;\lambda;\mu}\xi^{(2)\lambda} + \xi^{(1)}_{\nu;\lambda}\xi^{(2)\lambda}{}_{;\mu} \\
&\quad - \xi^{(2)}_{\nu;\lambda;\mu}\xi^{(1)\lambda} - \xi^{(2)}_{\nu;\lambda}\xi^{(1)\lambda}{}_{;\mu}. \tag{1.58}
\end{aligned}$$

From the fact that $\xi^{(1)\mu}$, $\xi^{(2)\mu}$ are Killing vectors, we have

$$\xi^{(i)}_{\mu;\nu;\lambda} + \xi^{(i)}_{\nu;\mu;\lambda} = 0, \quad i = 1, 2, \tag{1.59}$$

by taking the covariant derivative of Killing's equation. Also, from (1.26) we find that

$$\xi^{(i)}_{\mu;\nu;\lambda} = \xi^{(i)}_{\mu;\lambda;\nu} + \xi^{(i)\sigma} R_{\sigma\mu\nu\lambda}, \quad i = 1, 2. \tag{1.60}$$

With the use of (1.59) and (1.60) one can show that

$$\xi^{(1)}_{\mu;\lambda;\nu} \xi^{(2)\lambda} + \xi^{(1)}_{\nu;\lambda;\mu} \xi^{(2)\lambda} = \xi^{(1)\sigma} \xi^{(2)\lambda} (R_{\sigma\mu\lambda\nu} + R_{\sigma\nu\lambda\mu}), \tag{1.61a}$$

$$\xi^{(2)}_{\mu;\lambda;\nu} \xi^{(1)\lambda} + \xi^{(2)}_{\nu;\lambda;\mu} \xi^{(1)\lambda} = \xi^{(2)\sigma} \xi^{(1)\lambda} (R_{\sigma\mu\lambda\nu} + R_{\sigma\nu\lambda\mu}). \tag{1.61b}$$

Subtracting (1.61b) from (1.61a) we get

$$(\xi^{(1)}_{\mu;\lambda;\nu} + \xi^{(1)}_{\nu;\lambda;\mu})\xi^{(2)\lambda} - (\xi^{(2)}_{\mu;\lambda;\nu} + \xi^{(2)}_{\nu;\lambda;\mu})\xi^{(1)\lambda}$$
$$= \xi^{(1)\sigma} \xi^{(2)\lambda} (R_{\sigma\mu\lambda\nu} + R_{\sigma\nu\lambda\mu} - R_{\lambda\mu\sigma\nu} - R_{\lambda\nu\sigma\mu}) = 0, \tag{1.62}$$

where the last step follows from the symmetry properties of the Riemann tensor. Thus the terms on the right hand side of (1.58) involving double covariant derivatives vanish. The other terms can be shown to cancel by using Killing's equation. For example,

$$\begin{aligned}\xi^{(1)}_{\mu;\lambda} \xi^{(2)\lambda}{}_{;\nu} &= -\xi^{(1)}_{\lambda;\mu} \xi^{(2)\lambda}{}_{;\nu}\\ &= -\xi^{(1)\lambda}{}_{;\mu}\xi^{(2)}_{\lambda;\nu}\\ &= +\xi^{(1)\lambda}{}_{;\mu}\xi^{(2)}_{\nu;\lambda} \end{aligned} \tag{1.63}$$

which cancels the last term in (1.58), and so on. Thus ζ^μ satisfies (1.57) and so is a Killing vector. Suppose we have only n linearly independent Killing vectors $\xi^{(i)\mu}, i = 1, 2, \ldots, n$ and no more. Then the commutator of any two of these is a Killing vector and so must be a linear combination of some or all of the n Killing vectors with constant coefficients since there are no other solutions of Killing's equation. Thus we have the result

$$\xi^{(i)\mu}{}_{;\nu}\xi^{(j)\nu} - \xi^{(j)\mu}{}_{;\nu}\xi^{(i)\nu} = \sum_{k=1}^{n} a_k^{ij}\xi^{(k)\mu}, \quad i,j = 1,\ldots,n. \tag{1.64}$$

In coordinate independent notation, we can write

$$[\xi^{(i)}, \xi^{(j)}] = \sum_{k=1}^{n} a_k^{ij}\xi^{(k)}, \quad i,j = 1,\ldots,n. \tag{1.65}$$

In these two equations a_k^{ij} are constants.

1.5. Axially symmetric stationary metrics

Before we derive the most general axially symmetric stationary metric with the use of Killing vectors, we will derive it less rigorously but more simply from considerations of suitable coordinate systems and some reasonable physical assumptions. We will assume the field to be generated by the steady (that is, time-independent) rotation of a star made of perfect fluid, whose energy-momentum tensor is given by (1.36). The last is not a necessary assumption but helps to fix ideas. The star and the field around it possess axial symmetry about the axis of rotation which passes through the centre of the star, which we take to be the origin of the coordinate system. The axis of rotation is the z-axis. These last two sentences contain plausibility arguments because the geometry in the interior and the neighbourhood of the star is non-Euclidean and we cannot necessarily describe it in the simple terms that we are used to in Newtonian gravitation. Nevertheless because of the time independence and axial symmetry it would be reasonable to assume the existence of a time-like coordinate $x^0 = t$ and an angular coordinate $x^3 = \phi$ respectively of which the metric coefficients and all the matter variables are independent. If the reader prefers he or she can define time independence and axial symmetry by this last requirement. Thus we have

$$g_{\mu\nu} = g_{\mu\nu}(x^1, x^2), \quad \varepsilon = \varepsilon(x^1, x^2), \quad p = p(x^1, x^2). \tag{1.66}$$

where ε is the total mass-energy density and p is the pressure. Since ϕ is the angular coordinate about the rotation axis, the coordinate values (t, x^1, x^2, ϕ) and $(t, x^1, x^2, \phi + 2\pi)$ correspond to the same point in space–time:

$$(t, x^1, x^2, \phi) = (t, x^1, x^2, \phi + 2\pi). \tag{1.67}$$

The star's matter rotates in the ϕ direction, so its four-velocity $u^\mu = dx^\mu/ds$ has the following form:

$$u^0 = \frac{dt}{ds} = u^0(x^1, x^2), \quad u^1 = \frac{dx^1}{ds} = 0, \quad u^2 = \frac{dx^2}{ds} = 0,$$

$$u^3 = \frac{d\theta}{ds} = \frac{d\theta}{dt}\frac{dt}{ds} = \Omega(x^1, x^2)u^0. \tag{1.68}$$

Here $\Omega(x^1, x^2)$ is the angular velocity measured in units of coordinate time t. Equation (1.68) reflects the fact that a material particle in the star corresponds to fixed values of the coordinates x^1 and x^2 and only its ϕ coordinate changes with time. For rigid rotation Ω is constant (independent of x^1 and x^2) but for differential rotation Ω is in general a function of x^1 and x^2.

Because the star rotates in the ϕ direction, the field generated by it is not invariant under time reversal $t \to - t$, since such a transformation would reverse the sense of rotation of the star, resulting in a different space–time geometry. Nor is the star's field invariant under the transformation $\phi \to$

$- \phi$, for this would also reverse the sense of rotation of the star. However, the field of the star *is* invariant under a simultaneous reversal of t and ϕ: $(t, \phi \rightarrow -t, -\phi)$. In this case the motion would be exactly the same as before. Because of these properties the metric coefficients $g_{01}, g_{02}, g_{13}, g_{23}$ must vanish because otherwise under $(t, \phi \rightarrow -t, -\phi)$, for example, the term $g_{01} \, dt \, dx^1$ would change sign (recall that g_{01} is a function of x^1 and x^2 only) and so the metric would not be invariant, as it should be, under this transformation. Thus we have:

$$g_{01} = g_{02} = g_{13} = g_{23} = 0, \qquad (1.69)$$

so that the metric can be written as

$$ds^2 = g_{00} \, dt^2 + 2g_{03} \, dt \, d\phi + g_{33} \, d\phi^2 + g_{AB} \, dx^A \, dx^B, \qquad (1.70)$$

where A, B are to be summed over values 1, 2. It is possible to carry out an arbitrary coordinate transformation of the coordinates (x^1, x^2) to (x'^1, x'^2) without changing the form of the metric:

$$x^1 = x^1(x'^1, x'^2), \quad x^2 = x^2(x'^1, x'^2). \qquad (1.71)$$

The metric (1.70) can also be derived rigorously with the use of Killing vectors and by applying theorems due to Papapetrou (1966) and Carter (1969, 1970). We will not describe this derivation in detail but mention only the salient features. (See also the review article by Thorne (1971), for both the derivations given here.)

We assume the field to be that due to a rotating star and make the same assumptions about it as before. Then the space–time of the star (both interior and exterior) is stationary and axially symmetric. These properties are defined by requiring the metric to possess two linearly independent Killing vectors ξ and η such that ξ is everywhere time-like and η is everywhere space-like. Further, the orbits of the vector field η are closed. As one moves to a great spatial distance from the star, the gravitational influence of the star becomes negligible and the metric tends to that of flat (Minkowski) space, as follows:

$$ds^2 = dt^2 - d\rho^2 - dz^2 - \rho^2 \, d\phi^2, \qquad (1.72)$$

where we have used cylindrical polar coordinates. The fact that the metric tends to the Minkowski metric at infinite spatial distance from the source is referred to as asymptotic flatness.

We now show that the Killing vectors ξ and η commute everywhere, that is,

$$[\xi, \eta] = 0. \qquad (1.73)$$

At spatial infinity, where the metric is given by (1.72), the vectors ξ and η

have the following form (with $(x^0, x^1, x^2, x^3) = (t, \rho, z, \phi)$):

$$(\xi^0, \xi^1, \xi^2, \xi^3) = (1, 0, 0, 0), (\eta^0, \eta^1, \eta^2, \eta^3) = (0, 0, 0, 1), \qquad (1.74a)$$

$$(\xi_0, \xi_1, \xi_2, \xi_3) = (1, 0, 0, 0), (\eta_0, \eta_1, \eta_2, \eta_3) = (0, 0, 0, -\rho^2). \qquad (1.74b)$$

This is because the vectors given by (1.74a, b) satisfy Killing's equation for the metric (1.72):

$$\xi_{\mu;\nu} + \xi_{\nu;\mu} = 0, \quad \eta_{\mu;\nu} + \eta_{\nu;\mu} = 0. \qquad (1.75)$$

Equations (1.75) can be proved readily by noting that the only non-vanishing Christoffel symbols for the metric (1.72) are the following:

$$\Gamma^1_{33} = -\rho, \quad \Gamma^3_{13} = \rho^{-1}. \qquad (1.76)$$

The vectors given by (1.74a, b) also satisfy the conditions that ξ is time-like and η space-like with closed orbits (the latter are circles with constant ρ and z). For the metric (1.72) there are of course other Killing vectors which need not concern us. For the vectors (1.74a, b) we clearly have

$$\zeta^\mu = \xi^\mu{}_{;\nu}\eta^\nu - \eta^\mu{}_{;\nu}\xi^\nu = \xi^\mu{}_{,\nu}\eta^\nu - \eta^\mu{}_{,\nu}\xi^\nu = 0. \qquad (1.77)$$

Thus the commutator of ξ and η vanishes at spatial infinity. We now assume that ξ and η are the only Killing vectors of the space–time of the star (when we consider the whole space–time and not just spatial infinity). Then by the result (1.65) we get

$$[\xi, \eta] = a\xi + b\eta, \qquad (1.78)$$

where a and b are constants. But we have seen that the commutator of ξ and η vanishes at spatial infinity, where, however, ξ and η do not vanish (they have the form (1.74a, b)). Thus we must have $a = b = 0$, which shows that ξ and η commute everywhere, not just at infinity.

It follows from a result in differential geometry that if ξ and η commute everywhere, one can introduce coordinates t and ϕ such that

$$\xi = \frac{\partial}{\partial t}, \quad \eta = \frac{\partial}{\partial \phi}, \qquad (1.79)$$

using notation introduced earlier. We will refer to the other two coordinates as ρ and z (so that the latter are coordinates not just at spatial infinity as in (1.72) but over all space–time). The t-lines (that is, the curves $\rho = $ constant, $z = $ constant, $\phi = $ constant) are time-like everywhere and the ϕ-lines (that is, the curves $t = $ constant, $\rho = $ constant, $z = $ constant) are space-like and closed. With the use of this coordinate system the Killing vectors ξ and η are given by (1.74a) (but the covariant components are no longer given by (1.74b)) everywhere and not just at spatial infinity. It follows from the discussion below (1.53) that

$$\frac{\partial g_{\mu\nu}}{\partial t} = 0, \quad \frac{\partial g_{\mu\nu}}{\partial \phi} = 0, \qquad (1.80)$$

so that all the metric functions $g_{\mu\nu}$ are functions of ρ and z only, just like we assumed in the earlier derivation (see (1.66)).

To proceed further one uses what is referred to as orthogonal transitivity (Carter, 1969). Stated simply this asserts the existence of a family of two-dimensional surfaces which are orthogonal to the surfaces got by varying (t, ϕ) and keeping the other coordinates constant. The latter surfaces are surfaces of transitivity of the group of motions associated with the Killing vectors ξ and η. The group of motions associated with ξ are the transformations of space–time into itself given by $(x^0, x^1, x^2, x^3) \rightarrow (x^0 + t', x^1, x^2, x^3)$ for some parameter t', whereas those associated with η are given by $(x^0, x^1, x^2, x^3) \rightarrow (x^0, x^1, x^2, x^3 + \phi')$, for some ϕ'. Orthogonal transitivity implies the existence of coordinates ρ and z such that the vector fields $\zeta = \partial/\partial\rho$ and $\chi = \partial/\partial z$ are each orthogonal to both the vector fields ξ and η. Recalling that $(x^0, x^1, x^2, x^3) = (t, \rho, z, \phi)$, we get (1.69) as before. For example, ξ and ζ being normal implies $g_{\mu\nu}\xi^\mu\zeta^\nu = 0$. But $\xi^\mu = (1, 0, 0, 0)$ and $\zeta^\mu = (0, 1, 0, 0)$, so that $g_{01} = 0$, and so on. Thus we get the metric (1.70), as before, with $\rho = x^1, z = x^2$. The metric coefficients g_{00}, g_{03}, g_{33} have invariant meaning in the sense that they are the scalar products constructed out of the vectors ξ and η as follows:

$$g_{00} = g_{\mu\nu}\xi^\mu\xi^\nu, \quad g_{03} = g_{\mu\nu}\xi^\mu\eta^\nu, \quad g_{33} = g_{\mu\nu}\eta^\mu\eta^\nu. \qquad (1.81)$$

Carter (1969, 1970) has shown that the metric (1.70) can be derived under more general assumptions than those given here. For example, it is not necessary to assume asymptotic flatness nor make the assumption that there are no other Killing vectors than ξ and η. For one of the most important axially symmetric stationary metrics known, namely the Kerr (1963) solution, which is of the form (1.70), it is not necessarily true that the field is that of a star made of perfect fluid. We shall also be dealing with cylindrically symmetric metrics of the form (1.70), which are not asymptotically flat. The assumptions here should be regarded as sufficient but not necessary.

1.6. Connection between metric and angular momentum of source

We will now consider the connection between the metric due to a bounded rotating source and the mass and angular momentum of the source. The precise connection in all its details has not yet been worked out. The results we will state here will be adequate for our purpose. This question is discussed in detail in the book by Misner, Thorne and Wheeler (1973, Chapters 19 and 20).

In this section we use cartesian-like spatial coordinates (x^1, x^2, x^3) in the sense that at spatial infinity where space is flat these coordinates become cartesian coordinates and the metric at spatial infinity is given as follows:

$$ds^2 = dt^2 - (dx^1)^2 - (dx^2)^2 - (dx^3)^2. \tag{1.82}$$

We assume that the source has a centre which is located at $x^1 = x^2 = x^3 = 0$, and write $r^2 = (x^1)^2 + (x^2)^2 + (x^3)^2$. The variable r gives a measure of the spatial distance from the centre in the sense that for large values of r one is at a large spatial distance from the source and so the gravitational field is weak. It gives the exact spatial distance from the centre only if space is flat everywhere, which is not the case in the presence of a source. The metric has the following form for large values of r, that is, at a great spatial distance from the source (see Misner, Thorne and Wheeler, 1973, Chapter 19);

$$ds^2 = \left(1 - \frac{2M}{r} + A_0\right) dt^2 - \left(4\varepsilon_{ijk}\frac{S^j x^k}{r^3} + A_i\right) dt\, dx^i$$

$$- \left[\left(1 + \frac{2M}{r}\right)\delta_{jk} + A_{jk}\right] dx^j dx^k, \tag{1.83}$$

where i, j, k take values 1, 2, 3 and repeated indices are to be summed over these values. In (1.83), A_0 is $O(r^{-2})$, the A_i are $O(r^{-3})$ and the A_{jk} contain gravitational wave terms which are $O(r^{-1})$. The constant M is the total mass-energy of the source and the three-vector S^j is its total angular momentum. The vector S^j is a three-vector in the asymptotic Lorentz frame where the line-element (1.82) is valid. When the internal gravity of the source is weak, linearized approximations to Einstein's equations are valid and M and S^j can be expressed as simple integrals over the energy-momentum tensor. For strong fields, which is the situation we shall be concerned with, these expressions are not valid. However, the constants M and S^j can be interpreted as mass and angular momentum of the source for strong fields as well. The mass M is given by the Keplerian orbit of a gravitationally bound test particle at a large distance from the source as follows:

$$M = \omega^2 a^3, \tag{1.84}$$

where ω is the angular velocity of the test particle and a is the semi-major axis of its elliptical orbit. The angular momentum S^j is measured by placing a gyroscope at rest in the far gravitational field of the source, by applying a force to its centre of mass to prevent it from falling. The gyroscope will then precess relative to the coordinate system at infinity given by (1.82). The angular velocity of precession Ω^j of the gyroscope will be related to S^j, in suitable units, as follows:

$$\Omega^j = \frac{1}{r^3}\left[-S^j + \frac{3(S^k x^k)x^j}{r^2} \right].$$ (1.85)

This phenomenon is related to 'inertial dragging' mentioned earlier which was first pointed out by Thirring and Lense, 1918 (see also Landau and Lifshitz, 1975, p. 335).

If we specialize the metric (1.83) to a stationary and axially symmetric one there will be no gravitational wave terms so that the A_{jk} will be $O(r^{-2})$ and we may render the metric manifestly axially symmetric and of the form (1.70) if we assume $S^1 = S^2 = 0$, transform the coordinates as follows:

$$x^1 = x = \rho \cos \phi, \quad x^2 = y = \rho \sin \phi, \quad x^3 = z,$$ (1.86)

and choose the functions A_i and A_{jk} in the following manner:

$$A_1 = -\rho^{-1} \sin \phi A', \quad A_2 = \rho^{-1} \cos \phi A',$$ (1.87a)

$$A_{11} = A \cos^2 \phi + A'' \sin^2 \phi, \quad A_{22} = A \sin^2 \phi + A'' \cos^2 \phi,$$ (1.87b)

$$A_{12} = (A - A'') \sin \phi \cos \phi,$$

$$A_{13} = B \cos \phi, \quad A_{23} = B \sin \phi, \quad A_{33} = C.$$ (1.87c)

Here A_0, A, A', A'', B, C are all functions of ρ and z and they are all $O(r^{-2})$ or smaller, with $r^2 = \rho^2 + z^2$. The resulting metric is as follows:

$$ds^2 = \left(1 - \frac{2M}{r} + A_0 \right)dt^2 - \left(\frac{4S\rho^2}{r^3} + A' \right)dt\, d\phi - \rho^2 A'' d\phi^2$$

$$- \left(1 + \frac{2M}{r} \right)(d\rho^2 + dz^2 + \rho^2 d\phi^2) - A\, d\rho^2 - 2B\, d\rho\, dz - C\, dz^2.$$

(1.88)

Equation (1.88) gives the form of the metric at large distances from rotating bodies and shows the connection of the metric with the total mass-energy M and total angular momentum S of the source.

1.7. Further simplification of the rotating metric

We write the metric (1.70) as follows:

$$ds^2 = f\, dt^2 - 2k\, dt\, d\phi - l\, d\phi^2 - A\, d\rho^2 - 2B\, d\rho\, dz - C\, dz^2,$$ (1.89)

where f, k, l, A, B and C are all functions of ρ and z (the A, B, C here are distinct from those in the last section). We carry out a coordinate transformation from (ρ, z) to (ρ', z') as follows:

$$\rho' = F(\rho, z), \quad z' = G(\rho, z), \tag{1.90}$$

and write

$$\frac{\partial F}{\partial \rho} \equiv F_1, \quad \frac{\partial F}{\partial z} \equiv F_2, \quad \text{etc.} \tag{1.91}$$

Taking differentials of (1.90) and solving for $d\rho$ and dz we get

$$d\rho = J^{-1}(G_2 \, d\rho' - F_2 \, dz'), \quad dz = J^{-1}(-G_1 \, d\rho' + F_1 \, dz'), \tag{1.92}$$

where J is the Jacobian of the transformation (1.90) given by

$$J = \frac{\partial(F, G)}{\partial(\rho, z)} = F_1 G_2 - F_2 G_1, \tag{1.93}$$

and is assumed to be non-zero, for if it is zero F would be a function of G and (1.90) would not be a proper transformation. Substituting in (1.89) from (1.92) we get

$$\begin{aligned}
ds^2 = f \, dt^2 &- 2k \, dt \, d\phi - l \, d\phi^2 - J^{-2}\{(AG_2^2 - 2BG_1 G_2 + CG_1^2) \, d\rho'^2 \\
&+ 2[-AG_2 F_2 + B(G_2 F_1 + G_1 F_2) - CG_1 F_1] \, d\rho' \, dz' \\
&+ (AF_2^2 - 2BF_1 F_2 + CF_1{}^2) \, dz'^2\}.
\end{aligned} \tag{1.94}$$

The functions F and G are so far arbitrary. Assume that A, B, C are given functions of ρ and z. We now require F and G to satisfy the following two coupled non-linear partial differential equations as functions of ρ and z:

$$AG_2^2 - 2BG_1 G_2 + CG_1^2 = AF_2^2 - 2BF_1 F_2 + CF_1^2, \tag{1.95a}$$

$$-AG_2 F_2 + B(G_2 F_1 + G_1 F_2) - CG_1 F_1 = 0. \tag{1.95a}$$

We assume that for the given A, B, C the system of equations (1.95a, b) has a non-trivial solution with $J \neq 0$. Then in the coordinates (ρ', z') the metric (1.94) has its coefficient of $d\rho'^2$ equal to its coefficient of dz'^2 and the coefficient of $d\rho' \, dz'$ vanishes. We can now drop the primes from ρ' and z' and write the new metric as follows (the f, k, l in the following are not the same functions as the f, k, l in (1.89)):

$$ds^2 = f \, dt^2 - 2k \, dt \, d\phi - l \, d\phi^2 - e^\mu(d\rho^2 + dz^2), \tag{1.96}$$

where f, k, l and μ are all functions of ρ and z. Equation (1.96) will be the standard form of the metric for most of the remainder of the book.

1.8. Christoffel symbols and geodesics for a rotating metric

We shall often require the Christoffel symbols for the metric (1.96). We give these in this section, worked out with the use of (1.23). The non-zero covariant and contravariant components of the metric tensor are as follows (with $(x^0, x^1, x^2, x^3) = (t, \rho, z, \phi)$):

$$g_{00} = f, \quad g_{03} = -k, \quad g_{11} = g_{22} = -e^{\mu}, \quad g_{33} = -l, \qquad (1.97a)$$
$$g^{00} = D^{-2}l, \quad g^{03} = -D^{-2}k, \quad g^{11} = g^{22} = -e^{-\mu},$$
$$g^{33} = -D^{-2}f, \quad D^2 \equiv fl + k^2. \qquad (1.97b)$$

The non-zero Christoffel symbols $\Gamma^{\lambda}_{\mu\nu}$ are given in the following equations, where we have grouped them according to the value of the index λ ($\partial f/\partial \rho \equiv f_{\rho}$, $\partial f/\partial z \equiv f_z$, etc.):

$$\left.\begin{array}{ll} \Gamma^0_{01} = \tfrac{1}{2}D^{-2}(lf_{\rho} + kk_{\rho}), & \Gamma^0_{02} = \tfrac{1}{2}D^{-2}(lf_z + kk_z), \\ \Gamma^0_{13} = \tfrac{1}{2}D^{-2}(kl_{\rho} - lk_{\rho}), & \Gamma^0_{23} = \tfrac{1}{2}D^{-2}(kl_z - lk_z), \end{array}\right\} \qquad (1.98a)$$

$$\left.\begin{array}{lll} \Gamma^1_{00} = \tfrac{1}{2}e^{-\mu}f_{\rho}, & \Gamma^1_{03} = -\tfrac{1}{2}e^{-\mu}k_{\rho}, & \Gamma^1_{11} = \tfrac{1}{2}\mu_{\rho}, \\ \Gamma^1_{12} = \tfrac{1}{2}\mu_z, \ \Gamma^1_{22} = -\tfrac{1}{2}\mu_{\rho}, \ \Gamma^1_{33} = -\tfrac{1}{2}e^{-\mu}l_{\rho}, \end{array}\right\} \qquad (1.98b)$$

$$\left.\begin{array}{lll} \Gamma^2_{00} = \tfrac{1}{2}e^{-\mu}f_z, & \Gamma^2_{03} = -\tfrac{1}{2}e^{-\mu}k_z, & \Gamma^2_{11} = -\tfrac{1}{2}\mu_z, \\ \Gamma^2_{12} = \tfrac{1}{2}\mu_{\rho}, & \Gamma^2_{22} = \tfrac{1}{2}\mu_z, & \Gamma^2_{33} = -\tfrac{1}{2}e^{-\mu}l_z, \end{array}\right\} \qquad (1.98c)$$

$$\left.\begin{array}{ll} \Gamma^3_{01} = \tfrac{1}{2}D^{-2}(fk_{\rho} - kf_{\rho}), & \Gamma^3_{02} = \tfrac{1}{2}D^{-2}(fk_z - kf_z), \\ \Gamma^3_{13} = \tfrac{1}{2}D^{-2}(fl_{\rho} + kk_{\rho}), & \Gamma^3_{23} = \tfrac{1}{2}D^{-2}(fl_z + kk_z). \end{array}\right\} \qquad (1.98d)$$

We will now consider geodesics in the metric (1.96), which we can write down with the use of (1.98a–d). From the discussion following (1.14) we saw that in the Newtonian situation there is no transverse force on a test particle in the gravitational field of a rotating source. We implied in Fig. 1.2 that in general relativity there would be such a force. With some assumptions, we proceed to show this with the use of the geodesic equations (1.39). The coordinates of the test particle are all functions of the proper time s along the particle world-line. The geodesic equations are four equations for the four unknowns dt/ds, $d\rho/ds$, dz/ds and $d\phi/ds$. By virtue of (1.96) these functions satisfy the following equation:

$$f\left(\frac{dt}{ds}\right)^2 - 2k\frac{dt}{ds}\frac{d\phi}{ds} - l\left(\frac{d\phi}{ds}\right)^2 - e^{\mu}\left[\left(\frac{d\rho}{ds}\right)^2 + \left(\frac{dz}{ds}\right)^2\right] = 1. \quad (1.99)$$

The geodesic equations (1.39) in the metric (1.96) can be written as follows:

$$\frac{d^2t}{ds^2} + D^{-2}\frac{dt}{ds}\left[(lf_{\rho} + kk_{\rho})\frac{d\rho}{ds} + (lf_z + kk_z)\frac{dz}{ds}\right]$$

$$+ D^{-2}\frac{d\phi}{ds}\left[(kl_{\rho} - lk_{\rho})\frac{d\rho}{ds} + (kl_z - lk_z)\frac{dz}{ds}\right] = 0, \quad (1.100a)$$

$$\frac{d^2\rho}{ds^2} + \tfrac{1}{2}e^{-\mu}f_{\rho}\left(\frac{dt}{ds}\right)^2 - e^{-\mu}k_{\rho}\frac{dt}{ds}\frac{d\phi}{ds} + \tfrac{1}{2}\mu_{\rho}\left(\frac{d\rho}{ds}\right)^2$$

$$+ \mu_z \frac{d\rho\, dz}{ds\, ds} - \tfrac{1}{2}\mu_\rho \left(\frac{dz}{ds} \right)^2 - \tfrac{1}{2}e^{-\mu}l_\rho \left(\frac{d\phi}{ds} \right)^2 = 0, \quad (1.100b)$$

$$\frac{d^2 z}{ds^2} + \tfrac{1}{2}e^{-\mu}f_z \left(\frac{dt}{ds} \right)^2 - e^{-\mu}k_z \frac{dt\, d\phi}{ds\, ds} - \tfrac{1}{2}\mu_z \left(\frac{d\rho}{ds} \right)^2$$

$$+ \mu_\rho \frac{d\rho\, dz}{ds\, ds} + \tfrac{1}{2}\mu_z \left(\frac{dz}{ds} \right)^2 - \tfrac{1}{2}e^{-\mu}l_z \left(\frac{d\phi}{ds} \right)^2 = 0, \quad (1.100c)$$

$$\frac{d^2 \phi}{ds^2} + D^{-2}\frac{dt}{ds} \left[(fk_\rho - kf_\rho)\frac{d\rho}{ds} + (fk_z - kf_z)\frac{dz}{ds} \right]$$

$$+ D^{-2}\frac{d\phi}{ds} \left[(kk_\rho + fl_\rho)\frac{d\rho}{ds} + (kk_z + fl_z)\frac{dz}{ds} \right] = 0. \quad (1.100d)$$

Equation (1.99) is a first integral of (1.100a–d) so that one of the latter can be ignored in favour of (1.99). To follow the straight dashed path in Fig. 1.2, the path of the test particle must satisfy $d\phi/ds = 0$, since the ϕ coordinate does not change along this path. This would have to be true for any path (not necessarily confined to the equatorial plane) if the particle started from rest or had an initial velocity with no ϕ component, that is, if it were true that in general relativity there is no transverse force. We shall now show that for the test particle it is not possible to have a path with $d\phi/ds = 0$, unless the source is non-rotating. Putting $d\phi/ds = 0$ in (1.100d) (clearly in general we must have $dt/ds \neq 0$, $d\rho/ds \neq 0$, $dz/ds \neq 0$) we get

$$(fk_\rho - kf_\rho)\frac{d\rho}{ds} + (fk_z - kf_z)\frac{dz}{ds} = 0, \quad (1.101)$$

which after division by k^2, can be written as

$$-\frac{d}{ds}\left(\frac{f}{k} \right) = 0, \quad (1.102)$$

leading to $f = ak$, where a is a constant. If we now introduce a new time coordinate t' by the transformation $t' = t - a\phi$, it is easy to see that the coefficient of $dt'\, d\phi$ in the new metric vanishes. Such a metric is static, that is, it is produced by sources suffering no motion of any kind, so that the source is non-rotating. Referring to the metrics (1.83) and (1.88), we see that such a source would not have any angular momentum and hence would be non-rotating. An invariant characterization of this metric is that the time-like Killing vector is hypersurface orthogonal, that is, it is proportional to the gradient of a scalar. An equivalent condition is that the vector defined by (ξ^μ being the time-like Killing vector given by (1.74a))

$$\omega^\mu = \varepsilon^{\mu\nu\lambda\sigma}\xi_{\nu;\lambda}\xi_\sigma \quad (1.103)$$

24   *Introduction*

vanishes. Here $\varepsilon^{\mu\nu\lambda\sigma}$ is the Levi–Civita alternating tensor which is antisymmetric in any pair of indices with $\varepsilon^{0123} = (-g)^{-1/2}$. It is readily verified that ω^0 and ω^3 given by (1.103) vanish identically, and that the vanishing of ω^1 and ω^2 implies the following equations:

$$kf_\rho - fk_\rho = 0, \quad kf_z - fk_z = 0, \tag{1.104}$$

which again lead to $f = ak$. For further discussion of geodesics in a stationary gravitational field, the reader is referred to Landau and Lifshitz (1975, p. 252).[†]

We now write down the geodesic equations in the equatorial plane of a rotating source with reflection symmetry with respect to this plane, which we take to be $z = 0$. For the same reason as (1.15) all the z-derivatives vanish for $z = 0$. From (1.99) and (1.100a, d) we get, by putting $z = 0$,

$$f\left(\frac{dt}{ds}\right)^2 - 2k\frac{dt}{ds}\frac{d\phi}{ds} - l\left(\frac{d\phi}{ds}\right)^2 - e^\mu\left(\frac{d\rho}{ds}\right)^2 = 1, \tag{1.105a}$$

$$\frac{d^2t}{ds^2} + D^{-2}\frac{d\rho}{ds}\left[(lf_\rho + kk_\rho)\frac{dt}{ds} + (kl_\rho - lk_\rho)\frac{d\phi}{ds}\right] = 0, \tag{1.105b}$$

$$\frac{d^2\phi}{ds^2} + D^{-2}\frac{d\rho}{ds}\left[(fk_\rho - kf_\rho)\frac{dt}{ds} + (kk_\rho + fl_\rho)\frac{d\phi}{ds}\right] = 0. \tag{1.105c}$$

Here we have ignored (1.100b) in favour of (1.99). Equation (1.100c) is satisfied identically for $z = 0$. We shall use these equations in Section (4.4) for a rigidly rotating dust cylinder, since these equations are also valid on any plane $z =$ constant for a cylindrically symmetric system.

[†]There is a hiatus in this analysis in the sense that if ϕ is a periodic coordinate then the new time coordinate t' is also periodic; however, the analysis does indicate a connection between the transverse velocity $d\phi/ds$ and the metric function k, which is connected with the angular momentum.

2

The Einstein equations for a rotating metric and some classes of solutions

2.1. Introduction

The Einstein equations (1.35) in their generality are exceedingly complicated. To simplify these equations one usually sets $T_{\mu\nu} = 0$, so that one is considering the exterior, vacuum equations, and secondly one reduces the number of unknown functions and independent variables by using space–time symmetries, as we have done for the stationary axially symmetric metrics in the last chapter. One sets $T_{\mu\nu} = 0$ for several reasons. Firstly, the interior equations with $T_{\mu\nu} \neq 0$ are usually much more complicated than the exterior equations with $T_{\mu\nu} = 0$. Secondly, for many physical situations the form of $T_{\mu\nu}$ is uncertain. Thirdly, a study of the solutions with $T_{\mu\nu} = 0$, that is, the exterior solutions, gives an idea as to the sources these solutions might represent, whereas the corresponding interior solutions might be difficult or impossible to find like the interior solution for the Kerr solution. For these reasons we shall firstly concentrate on studying the exterior Einstein equations for the rotating metric (1.96).

The vacuum Einstein equations are given by

$$R_{\mu\nu} = 0, \qquad (2.1)$$

which is obtained by setting $T_{\mu\nu} = 0$ in (1.35) and using the fact that in this case $R = 0$.

2.2. Exterior field equations

We label the coordinates as before: $(x^0, x^1, x^2, x^3,) = (t, \rho, z, \phi)$. Three of the equations (2.1) can be written as follows:

$$2e^\mu D^{-1} R_{00} = (D^{-1} f_\rho)_\rho + (D^{-1} f_z)_z + D^{-3} f(f_\rho l_\rho + f_z l_z + k_\rho^2 + k_z^2) = 0, \qquad (2.2a)$$

$$-2e^\mu D^{-1} R_{03} = (D^{-1} k_\rho)_\rho + (D^{-1} k_z)_z + D^{-3} k(f_\rho l_\rho + f_z l_z + k_\rho^2 + k_z^2) = 0, \qquad (2.2b)$$

$$-2e^\mu D^{-1}R_{33} = (D^{-1}l_\rho)_\rho + (D^{-1}l_z)_z + D^{-3}l(f_\rho l_\rho + f_z l_z + k_\rho^2 + k_z^2) = 0$$
$$(2.2c)$$

From $(2.2a, b, c)$ one gets the following equation:

$$e^\mu D^{-1}(lR_{00} - 2kR_{03} - fR_{33}) = D_{\rho\rho} + D_{zz} = 0. \tag{2.3}$$

Thus the function D satisfies the two-dimensional Laplace equation in the variables ρ and z. It follows that D can be considered as the real part of an analytic function $\Sigma(\rho + iz)$ of $\rho + iz$. Let E be the imaginary part of $\Sigma(\rho + iz)$, that is,

$$\Sigma(\rho + iz) = D(\rho, z) + iE(\rho, z). \tag{2.4}$$

The Cauchy–Riemann equations imply

$$D_\rho = E_z, \quad D_z = -E_\rho. \tag{2.5}$$

Consider now the transformation from (ρ, z) to $(\bar\rho, \bar z)$ given by

$$\bar\rho = D(\rho, z), \quad \bar z = E(\rho, z). \tag{2.6}$$

Because of (2.5) we have

$$(d\bar\rho)^2 + (d\bar z)^2 = (D_\rho^2 + E_\rho^2)(d\rho^2 + dz^2). \tag{2.7}$$

Equation (2.7) shows that the form of the metric (1.96) is unaltered by the transformation (2.6), since we can define a new function $\bar\mu$ given by

$$e^{\bar\mu} = e^\mu (D_\rho^2 + E_\rho^2)^{-1}. \tag{2.8}$$

We can assume that all the functions $f, k, l, \bar\mu$ have been expressed in terms of the variables $(\bar\rho, \bar z)$ obtained by substituting for (ρ, z) after solving for the latter from (2.6). Having expressed all functions in terms of $(\bar\rho, \bar z)$, we can drop the bars so that because of (2.6) we are left with the following algebraic relation in f, k and l:

$$D^2 = fl + k^2 = \rho^2. \tag{2.9}$$

The above procedure was first used by Weyl (1917) for the axially symmetric static metric (with $k = 0$) and generalized to the present case by Lewis (1932). We drop the bar from $\bar\mu$ as well, so that the rest of the non-trivial Einstein equations (with the use of (2.9)) can be written as follows:

$$2R_{11} = -\mu_{\rho\rho} - \mu_{zz} + \rho^{-1}\mu_\rho + \rho^{-2}(f_\rho l_\rho + k_\rho^2) = 0, \tag{2.10a}$$
$$2R_{12} = \rho^{-1}\mu_z + \tfrac{1}{2}\rho^{-2}(f_\rho l_z + f_z l_\rho + 2k_\rho k_z) = 0, \tag{2.10b}$$
$$2R_{22} = -\mu_{\rho\rho} - \mu_{zz} - \rho^{-1}\mu_\rho + \rho^{-2}(f_z l_z + k_z^2) = 0. \tag{2.10c}$$

Because of (2.9) only two of $(2.2a, b, c)$ are independent. It is more convenient to use the function w instead of k defined by

$$w = f^{-1}k. \tag{2.11}$$

From (2.2a, b) eliminating k and l we get the following two equations:

$$f(f_{\rho\rho} + f_{zz} + \rho^{-1}f_{\rho}) - f_{\rho}^2 - f_z^2 + \rho^{-2}f^4(w_{\rho}^2 + w_z^2) = 0, \quad (2.12a)$$

$$f(w_{\rho\rho} + w_{zz} - \rho^{-1}w_{\rho}) + 2f_{\rho}w_{\rho} + 2f_z w_z = 0. \quad (2.12b)$$

From (2.10a, b, c) we can express μ_{ρ} and μ_z in terms of f and w as follows:

$$\mu_{\rho} = -f^{-1}f_{\rho} + \tfrac{1}{2}\rho f^{-2}(f_{\rho}^2 - f_z^2) - \tfrac{1}{2}\rho^{-1}f^2(w_{\rho}^2 - w_z^2), \quad (2.13a)$$

$$\mu_z = -f^{-1}f_z + \rho f^{-2}f_{\rho}f_z - \rho^{-1}f^2 w_{\rho}w_z, \quad (2.13b)$$

the consistency of which is guaranteed by (2.12a, b). This is seen as follows. Equations (2.13a, b) are consistent if the z-derivative of the right hand side of (2.13a) is equal to the ρ-derivative of the right hand side of (2.13b). This condition is given by

$$\rho f^{-3}f_z[f(f_{\rho\rho} + f_{zz} + \rho^{-1}f_{\rho}) - f_{\rho}^2 - f_z^2]$$
$$+ \rho^{-1}f^2 w_z(w_{zz} + w_{\rho\rho}) - \rho^{-1}ff_z(w_{\rho}^2 - w_z^2)$$
$$- \rho^{-2}f^2 w_{\rho}w_z + 2\rho^{-1}ff_{\rho}w_{\rho}w_z = 0. \quad (2.13)$$

The terms in equation (2.13) which depend on f only can be replaced with the use of (2.12a) by the expression $\rho^{-1}ff_z(w_{\rho}^2 + w_z^2)$. We carry out this replacement in (2.13) and then divide by $\rho^{-1}fw_z$. The resulting equation is (2.12b). Thus the basic equations are (2.12a, b). Once these equations are solved for f and w, the resulting μ can be got from (2.13a, b) by quadratures, that is, by a simple integration. The function μ, however, satisfies the three equations (2.10a, b, c). We still have to verify that the μ obtained from (2.13a, b) satisfies either (2.10a) or (2.10c) (since (2.13a) is derived from these two equations, it is sufficient to verify that one of (2.10a, c) is satisfied by μ given by (2.13a, b)). We shall show that (2.10a) is satisfied.

Eliminating l and k from (2.10a) this equation can be written as follows:

$$\mu_{\rho\rho} + \mu_{zz} = \rho^{-1}f^{-1}f_{\rho} - \tfrac{1}{2}f^{-2}(f_{\rho}^2 + f_z^2) + \tfrac{1}{2}\rho^{-2}f^2(w_{\rho}^2 + w_z^2). \quad (2.14)$$

Adding the ρ-derivative of (2.13a) to the z-derivative of (2.13b) we get

$$\mu_{\rho\rho} + \mu_{zz} = \tfrac{3}{2}f^{-2}f_{\rho}^2 + \tfrac{1}{2}f^{-2}f_z^2 - \rho f^{-3}f_{\rho}(f_{\rho}^2 + f_z^2) - f^{-1}(f_{\rho\rho} + f_{zz})$$
$$+ \rho f^{-2}f_{\rho}(f_{\rho\rho} + f_{zz}) + \tfrac{1}{2}\rho^{-2}f^2(w_{\rho}^2 - w_z^2) - \rho^{-1}ff_{\rho}(w_{\rho}^2 - w_z^2)$$
$$- \rho^{-1}f^2 w_{\rho}(w_{\rho\rho} + w_{zz}) - 2\rho^{-1}ff_z w_{\rho}w_z. \quad (2.15)$$

If we now replace the second derivatives in (2.15) by the corresponding expressions from (2.12a, b) we find, after some manipulation. that (2.15) reduces to (2.14).

With the use of the relations (2.9) and (2.11) the metric (1.96) is sometimes written as follows:

$$ds^2 = f(dt - wd\phi)^2 - \rho^2 f^{-1}d\phi^2 - e^{\mu}(d\rho^2 + dz^2). \quad (2.16)$$

We call this the Weyl–Lewis–Papapetrou form of the metric. With the use of a different approach to the Einstein equations for stationary axisymmetric fields, Cosgrove (1978) has taken the function

$$\gamma = \tfrac{1}{2}(\mu + \log f) \qquad (2.17)$$

as the basic unknown instead of f and w as in the above formulation. He obtains a single fourth order quasi-linear (that is, the fourth-derivative terms are linear but the others are not) partial differential equation for γ. However, once this equation has been solved, the other functions f and w are not given by quadratures but one has to solve a pair of second order ordinary differential equations. The consequences of this approach have not been fully worked out and this could turn out to be an interesting formulation.

2.3. The Weyl solutions

When there is no rotation, that is, when $w = 0$, (2.12a) can be solved by putting $f = e^{\sigma}$, where σ is harmonic, that is, it satisfies the Laplace equation:

$$\nabla^2 \sigma = \sigma_{\rho\rho} + \sigma_{zz} + \rho^{-1}\sigma_{\rho} = 0. \qquad (2.18)$$

The resulting metric can be written as follows:

$$ds^2 = e^{\sigma}\,dt^2 - e^{-\sigma}[e^{\chi}(d\rho^2 + dz^2) + \rho^2\,d\phi^2], \qquad (2.19)$$

where the function χ is given by the following two equations:

$$\chi_z = \rho\sigma_{\rho}\sigma_z, \quad \chi_{\rho} = \tfrac{1}{2}\rho(\sigma_{\rho}^2 - \sigma_z^2), \qquad (2.20)$$

the consistency of which is guaranteed by (2.18). Equations (2.18)–(2.20) give the Weyl (1917) solutions which represent axially symmetric static (non-rotating) exterior fields. A particular class of Weyl solutions of interest is given by the following harmonic σ:

$$\sigma = \delta \log\left(\frac{z - m + R^{(-)}}{z + m + R^{(+)}}\right) = \delta \log\left(\frac{R^{(-)} + R^{(+)} - 2m}{R^{(-)} + R^{(+)} + 2m}\right),$$

$$R^{(\pm)} = [\rho^2 + (z \pm m)^2]^{1/2}, \qquad (2.21)$$

where δ and m are constants. The function σ given by (2.21) is the external Newtonian gravitational potential of an infinitesimally thin uniform rod with density proportional to δ, with its centre at the origin of coordinates and its ends lying on the z-axis at $z = -m$ and $z = +m$, so that it has length $2m$. This can be seen by evaluating the integral (1.4) with $\varepsilon = \varepsilon_0\delta(x')\delta(y')$, where $\delta(x)$ is the Dirac delta function and ε_0 is a suitable constant, and the z' integral ranges from $z' = -m$ to $z' = +m$. However, the source of the corresponding metric (2.19) in general relativity is not necessarily a uniform rod, as can be seen from the fact that for $\delta = 1$ (2.21) leads to the Schwarzschild metric, which is spherically symmetric (see the next section). For $\delta = 2, 3, \ldots$ (2.21) leads to the static limit of the Tomimatsu–Sato solutions considered in the next chapter.

2.4. The Schwarzschild solution

We shall not be concerned with the Schwarzschild solution except incidentally, mainly as the limit of the Kerr solution discussed in the next chapter. The Schwarzschild solution is very important but it is discussed in all standard text books on general relativity and we shall not discuss it in detail. We refer to Misner, Thorne and Wheeler (1973, Chapter 23) for more details of the material in this section.

As mentioned earlier, the Weyl solution given by (2.19), (2.20) and (2.21) with $\delta = 1$ leads to the Schwarzschild solution, the standard form of which is

$$ds^2 = \left(1 - \frac{2m}{r}\right) dt^2 - \left(1 - \frac{2m}{r}\right)^{-1} dr^2 - r^2 (d\theta^2 + \sin^2\theta \, d\phi^2). \quad (2.22)$$

The transformation which takes the Weyl form of the Schwarzschild solution to (2.22) is as follows:

$$\rho = (r^2 - 2mr)^{1/2} \sin\theta, \quad z = (r - m)\cos\theta, \quad (2.23a)$$

$$r = m + \tfrac{1}{2}(R^{(+)} + R^{(-)}), \quad \cos\theta = \frac{1}{2m}(R^{(+)} - R^{(-)}), \quad (2.23b)$$

where $R^{(\pm)}$ are given by (2.21). We leave the derivation of (2.22) from the Weyl form as an exercise for the reader. The standard way of deriving the Schwarzschild metric is different and is given in most text books.

The Schwarzschild solution is spherically symmetric, as is seen by the fact that (2.22) is invariant under an arbitrary rotation of the coordinate system considered as a spherical polar coordinate system. The metric (2.22) represents the exterior field of a spherically symmetric distribution of matter of mass m, but, as is well known, it can also be interpreted as the field of a highly collapsed spherically symmetric star, that is, a black hole.

The metric (2.22) is regular, that is, all the metric functions $g_{\mu\nu}$, the functions representing the contravariant components $g^{\mu\nu}$ and their partial derivatives are continuous everywhere except at $r = 0$ and $r = 2m$. The latter, however, is only a coordinate irregularity which can be removed by Kruskal's (1960) transformation to new coordinates (u, v) from (r, t) given implicitly by

$$u^2 - v^2 = \frac{1}{2m}(r - 2m)e^{r/2m}, \quad v = u \tanh(t/4m), \quad (2.24)$$

which reduces (2.22) to the following form

$$ds^2 = \frac{32m^3}{r}e^{-r/2m}(du^2 - dv^2) + r^2(d\theta^2 + \sin^2\theta d\phi^2). \qquad (2.25)$$

Here r is to be interpreted as a function of u and v given by the first equation in (2.24). The irregular behaviour at $r = 0$, however, cannot be removed by any coordinate transformation. One way to see this is to compute the scalar $R_{\lambda\mu\nu\sigma}R^{\lambda\mu\nu\sigma}$, known as a curvature scalar, which tends to infinity at $r = 0$. Since a scalar has values which are independent of the coordinate system, this scalar will tend to infinity at the point given by $r = 0$ in any coordinate system. Such a point is called a space–time singularity. If a test particle is dropped, say from rest, at any finite value of r, it will reach the point $r = 0$ in a finite proper time, that is, time measured by an observer on the test particle. Moreover, it can be shown by considering the equation for geodesic deviation (this is the equation which measures the rate at which neighbouring geodesics converge or diverge) that as any material particle nears the point $r = 0$, it will be disrupted by tidal gravitational forces which tend to infinity as r tends to zero.

Although the surface $r = 2m$ is not a space–time singularity, it has physical significance. This surface is a 'one-way membrane' in the sense that light-like and time-like geodesics go into the region $r < 2m$ from the region $r > 2m$ but no light-like and time-like geodesics come out of the region $r < 2m$. Thus no light signals or material particles can come out of the region $r < 2m$. We assume here that the mass m is very much larger than those of the test particles so that the latter do not effect the geometry. Thus the space–time singularity at $r = 0$ is hidden from the region $r > 2m$ by the surface $r = 2m$, which is called an *event horizon* or just a *horizon*.

2.5. The Papapetrou solutions

Although the Papapetrou solutions of equations (2.12a, b) are physically not very interesting we shall consider these because they form one of the few classes of solutions which depend on a harmonic function, and the derivation of these solutions gives some feeling for the equations and illustrates the difficulty of finding physically interesting solutions. These solutions also provide the basis for an approximate class of solutions of some physical interest which we will consider below.

We start with the assumption

$$w_{\rho\rho} + w_{zz} - \rho^{-1}w_\rho = 0. \qquad (2.26)$$

The solution of (2.26) is given in terms of a harmonic function ζ as follows:

$$w = A\rho\zeta_\rho, \quad \nabla^2\zeta = \zeta_{\rho\rho} + \zeta_{zz} + \rho^{-1}\zeta_\rho = 0, \tag{2.27}$$

where A is an arbitrary constant. Equations (2.12b) and (2.26) imply

$$f_\rho w_\rho + f_z w_z = 0, \tag{2.28}$$

which, with the use of (2.27), can be written as

$$- f_\rho\zeta_{zz} + f_z\zeta_{\rho z} = 0, \tag{2.29}$$

the general solution of which is

$$f = f(\zeta_z), \tag{2.30}$$

that is, f is an arbitrary function of ζ_z. From this equation we get

$$f_\rho = f'\zeta_{\rho z}, \quad f_{\rho\rho} = f''\zeta_{\rho z}^2 + f'\zeta_{\rho\rho z}, \tag{2.31a}$$

$$f_z = f'\zeta_{zz}, \quad f_{zz} = f''\zeta_{zz}^2 + f'\zeta_{zzz} \tag{2.31b}$$

where $f' = \partial f/\partial\zeta_z$. With the use of (2.27), (2.31a, b) and the relation

$$\zeta_{\rho\rho z} + \zeta_{zzz} + \rho^{-1}\zeta_{\rho z} = 0, \tag{2.32}$$

which follows from (2.27), equation (2.12a) reduces to the form

$$ff'' - f'^2 + A^2 f^4 = 0. \tag{2.33}$$

After the substitution $f = h^{-1}$, this equation can be solved as follows:

$$h = f^{-1} = \alpha\cosh\zeta_z + \beta\sinh\zeta_z, \quad A^2 = \alpha^2 - \beta^2, \tag{2.34}$$

where α and β are arbitrary constants related to A as in (2.34). Thus (2.27) and (2.34) give a class of exact solutions of (2.12a, b) in terms of the harmonic function ζ. This is the Papapetrou (1953) class of solutions.

To get an asymptotically flat solution we can, for example, take ζ to be the harmonic function $r^{-1}, r = (\rho^2 + z^2)^{1/2}$ (this is different from the r of the Schwarzschild solution). In this case

$$w = - A\rho^2 r^{-3}. \tag{2.35}$$

From (1.88) and (1.96) we see that for asymptotically flat solutions the form of f and k at infinity is

$$f = 1 - \frac{2M}{r} + \cdots, \quad k = -\frac{2S\rho^2}{r^3} + \cdots \tag{2.36}$$

where dots represent terms which vanish at infinity faster than the terms preceding the dots. Looking at the definition (2.11) of w we see from (2.36) that w has the same leading term as k in (2.36). Thus (2.35) gives the correct asymptotic form for w. However, the function f given by (2.34) with $\zeta = r^{-1}$

has the following form for large r:

$$f = \alpha^{-1}\left(1 + \frac{\beta}{\alpha}\frac{z}{r^3} + \cdots\right) \tag{2.37}$$

where dots represent terms which vanish faster than r^{-2}. Comparing with the form of f in (2.36) we see that the f given by (2.37) has no term proportional to r^{-1} so that it has zero mass. It can similarly be shown that any choice of ζ that gives the correct asymptotic form for w gives a corresponding f which has no mass term. This class of solutions, therefore, does not contain any solution representing a massive bounded rotating source whose field is asymptotically flat.

2.6. Lewis and Van Stockum solutions

In this section we shall consider two more related classes of solutions of (2.12*a, b*) depending on a harmonic function. These solutions cannot represent the exterior field of a bounded rotating source for they do not contain any asymptotically flat solutions. Nevertheless they are of some interest, firstly, because we shall be using the cylindrically symmetric form of these solutions in our study of rotating neutral dust in Chapter 4. Secondly, they are of some interest as regards generating new solutions of (2.12*a, b*). Thus, for example, the Van Stockum solutions have been used by Herlt (1978) to generate asymptotically flat static (non-rotating) solutions of the Einstein–Maxwell equations. Thirdly, the derivation of a class of solutions of (2.12*a, b*) depending on a harmonic function gives some insight into the structure of these equations even if this class contains no asymptotically flat solutions. Lewis found his class of solutions in 1932. Van Stockum (1937), in the process of examining the Lewis solutions, found the class bearing his name. Here we shall follow the derivation of Van Stockum. Tanabe (1979) was able to generate the Lewis solutions from the Weyl solutions.

Instead of tackling (2.12*a, b*), we revert to (2.2*a, b, c*). With the use of (2.9), we write (2.2*a, b, c*) as follows:

$$(\rho^{-1}f_\rho)_\rho + (\rho^{-1}f_z)_z + \rho^{-3}f\Sigma = 0, \tag{2.38a}$$

$$(\rho^{-1}k_\rho)_\rho + (\rho^{-1}k_z)_z + \rho^{-3}k\Sigma = 0, \tag{2.38b}$$

$$(\rho^{-1}l_\rho)_\rho + (\rho^{-1}l_z)_z + \rho^{-3}l\Sigma = 0, \tag{2.38c}$$

where

$$\Sigma \equiv f_\rho l_\rho + f_z l_z + k_\rho^2 + k_z^2. \tag{2.38d}$$

Recall that only two of (2.38*a, b, c*) are independent. Eliminating Σ between

(2.38*a, c*) and (2.38*b, c*) we get the following two equations:

$$[\rho^{-1}(fl_\rho - lf_\rho)]_\rho + [\rho^{-1}(fl_z - lf_z)]_z = 0, \qquad (2.39a)$$

$$[\rho^{-1}(kl_\rho - lk_\rho)]_z + [\rho^{-1}(kl_z - lk_z)]_z = 0. \qquad (2.39b)$$

We introduce the functions *u* and *v* as follows:

$$u = l^{-1}f, \quad v = l^{-1}k. \qquad (2.40)$$

In terms of *u* and *v* (2.39*a, b*) can be written as

$$[(u + v^2)^{-1}\rho u_\rho]_\rho + [(u + v^2)^{-1}\rho u_z]_z = 0, \qquad (2.41a)$$

$$[(u + v^2)^{-1}\rho v_\rho]_\rho + [(u + v^2)^{-1}\rho v_z]_z = 0. \qquad (2.41b)$$

We look for a solution of (2.41*a, b*) in terms of functions *U* and *V* given by

$$u_\rho = (u + v^2)U_\rho, \quad u_z = (u + v^2)U_z, \qquad (2.42a)$$

$$v_\rho = (u + v^2)V_\rho, \quad v_z = (u + v^2)V_z. \qquad (2.42b)$$

Equations (2.41*a, b*) and (2.42*a, b*) imply that *U* and *V* should satisfy

$$\nabla^2 U = U_{\rho\rho} + U_{zz} + \rho^{-1}U_\rho = 0, \qquad (2.43a)$$

$$\nabla^2 V = V_{\rho\rho} + V_{zz} + \rho^{-1}V_\rho = 0, \qquad (2.43b)$$

so that *U* and *V* are both harmonic functions. It is easily seen that (2.42*a, b*), (2.43*a, b*) are compatible if we take

$$v = Au + B, \quad V = AU + B, \qquad (2.44)$$

where *A* and *B* are constants. Equations (2.42*a, b*) and (2.44) then imply that μ is a function of *U* given by

$$\frac{du}{A^2u^2 + (2AB + 1)u + B^2} = dU. \qquad (2.45)$$

Three cases arise according as to whether $4AB + 1 < 0, = 0$ or > 0.

Case (i): $4AB + 1 = -4u_0^2 A^4 < 0$, where u_0 is a constant. Then (2.45) integrates to

$$u = u_0^2 A^2 - (4A^2)^{-1} + u_0 \tan U, \qquad (2.46a)$$

$$v = -(2A)^{-1} + Au_0 \tan U, \qquad (2.46b)$$

where for later convenience we have replaced $u_0 A^2 U$ by *U* since *U* is undefined to within a constant factor. From (2.9) and (2.40) we get

$$f = lu, \quad k = lv, \quad l = \rho(u + v^2)^{-1/2} \qquad (2.47)$$

From (2.46*a, b*) and (2.47) we get

$$l = (u_0 A)^{-1}\rho \cos U, \qquad (2.48a)$$

$$f = (u_0 A)^{-1}\rho[(u_0^2 A^2 - (4A^2)^{-1})\cos U + u_0 \sin U], \qquad (2.48b)$$

$$k = (u_0 A)^{-1}\rho[-(2A)^{-1}\cos U + Au_0 \sin U]. \qquad (2.48c)$$

From (2.10a, b, c) we get

$$\mu_\rho = \tfrac{1}{2}\rho^{-1}(f_z l_z + k_z^2 - f_\rho l_\rho - k_\rho^2), \tag{2.49a}$$

$$\mu_z = -\tfrac{1}{2}\rho^{-1}(f_\rho l_z + f_z l_\rho + 2k_\rho k_z). \tag{2.49b}$$

With the use of (2.48a, b, c) and (2.49a, b) we get

$$\mu_\rho = -\tfrac{1}{2}\rho^{-1} - \tfrac{1}{2}\rho(U_z^2 - U_z^2), \quad \mu_z = -\rho U_\rho U_z. \tag{2.50}$$

Case (ii): $4AB + 1 = 0$. In this case (2.45) integrates to

$$u = -(4A^2)^{-1} - (A^2 U)^{-1}, \tag{2.51a}$$

$$v = -(2A)^{-1} - (AU)^{-1}. \tag{2.51b}$$

From (2.47) and (2.51a, b) we then get

$$l = \rho A U, \tag{2.52a}$$

$$f = \rho(-(4A)^{-1}U - A^{-1}), \tag{2.52b}$$

$$k = \rho(-\tfrac{1}{2}U - 1). \tag{2.52c}$$

If we now carry out a transformation to a new time coordinate t' as follows (with b as a constant):

$$t' = t - b\phi, \tag{2.53}$$

the metric (1.96) becomes

$$ds^2 = f'dt'^2 - 2k'd\phi dt' - l'd\phi^2 - e^\mu(d\rho^2 + dz^2), \tag{2.54}$$

where

$$f' = f, \tag{2.54a}$$

$$k' = k - bf, \tag{2.54b}$$

$$l' = l + 2bk - b^2 f. \tag{2.54c}$$

With the choice $b = 2A$, we get, from (2.52a, b, c),

$$l' = 0, \tag{2.55a}$$

$$f' = \rho(-(4A)^{-1}U - A^{-1}), \tag{2.55b}$$

$$k' = \rho. \tag{2.55c}$$

Defining a new harmonic function $U' = -(4A)^{-1}U - A^{-1}$, the metric can now be written as

$$ds^2 = \rho U'dt'^2 - 2\rho d\phi dt' - \rho^{1/2}(d\rho^2 + dz^2), \tag{2.56}$$

where the function μ has been evaluated by substituting from (2.52a, b, c) into (2.49a, b). In fact (2.49a, b) are invariant under the transformation (2.54a, b, c) so we could equally well substitute the primed quantities (2.55a, b, c) into (2.49a, b) for the evaluation of μ. Equation (2.56) gives the

Van Stockum solution (see, for example, equation (18.23) of Kramer *et al.* (1980)). This is the Van Stockum *exterior* solution which is quite distinct from the Van Stockum *interior* (dust) solution which we will consider in Chapter 4.

Case (iii): $4AB + 1 = + 4u_0^2 A^4 > 0$, where u_0 is a constant. (2.45) integrates to

$$u = - u_0^2 A^2 - (4A^2)^{-1} + u_0 \coth U, \tag{2.57a}$$

$$v = - (2A)^{-1} + u_0 A \coth U, \tag{2.57b}$$

where we have written U for $- u_0 A^2 U$ since the latter is also harmonic. From (2.57a, b) and (2.47) we get

$$l = (u_0 A)^{-1} \rho \sinh U, \tag{2.58a}$$

$$f = (u_0 A)^{-1} \rho [(- u_0^2 A^2 - (4A^2)^{-1} \sinh U + u_0 \cosh U], \tag{2.58b}$$

$$k = (u_0 A)^{-1} \rho [- (2A)^{-1} \sinh U + u_0 A \cosh U]. \tag{2.58c}$$

From (2.49a, b) and (2.58a, b, c) we get instead of (2.50) the following relations for μ:

$$\mu_\rho = - \tfrac{1}{2} \rho^{-1} + \tfrac{1}{2} \rho (U_\rho^2 - U_z^2), \quad \mu_z = \rho U_\rho U_z. \tag{2.59}$$

Equations (2.48a, b, c), (2.50), (2.58a, b, c) and (2.59) give the Lewis solutions, the cylindrically symmetric form of which we will use in Chapter 4.

It is readily verified that the Lewis and Van Stockum solutions contain no asymptotically flat solutions. Recall that for such solutions the functions f and k must have the form (2.36) at infinity. Let us try to make (2.48a, b, c) conform to the behaviour (2.36). The simplest way to make k have the right behaviour at infinity is to transform to a new time coordinate t' given by (2.53), and choose b such that in the k' given by (2.54b) the $\cos U$ term cancels out. Then k' will have the correct behaviour at infinity if we choose U to vanish at infinity like ρr^{-3}, where $r^2 = \rho^2 + z^2$. However, even if this can be done, it is then seen that the corresponding f' (which is the same as f given by (2.48b)) will not have the behaviour in (2.36). In fact f will grow at infinity like ρ. Similar remarks apply to the other solutions found in this section.

The Lewis and Van Stockum solutions, could, in principle, represent the exterior field of a distribution of rotating matter which extends to infinity in both directions along the axis of symmetry. This distribution of matter need not be cylindrically symmetric, that is, it can have a z-dependence. A cylindrically symmetric form of this situation will be considered in Chapter 4.

2.7. A class of approximate solutions

We saw in Section 2.5 that the solutions in the Papapetrou class that are asymptotically flat have zero mass. The Weyl class of solutions given by (2.18)–(2.20) clearly contain asymptotically flat solutions with mass, but these are non-rotating. For example, if we choose the harmonic function σ of (2.18) as follows:

$$\sigma = -\frac{2m}{r}, \quad r^2 = \rho^2 + z^2, \tag{2.60}$$

then the resulting solution will be asymptotically flat and will have mass m (this is the Curzon solution (1924)). Thus if we could somehow combine the Weyl and Papapetrou solutions, we might get asymptotically flat solutions with non-zero mass. More specifically one might ask whether there exists a class of solutions of (2.12*a, b*) depending on two harmonic functions σ and ζ (see (2.18) and (2.27)) such that if $\zeta = 0$ one gets the Weyl class of solutions, but if $\sigma = 0$ one gets the Papapetrou class of solutions. In this section we shall derive an approximate class of solutions which has this property (see Islam (1976*a, c*)).

We first transform the equations (2.12*a, b*) by the substitution $f = h^{-1}$ to the following pair of equations:

$$h(h_{\rho\rho} + h_{zz} + \rho^{-1}h_{\rho}) - h_{\rho}^2 - h_z^2 - \rho^{-2}(w_{\rho}^2 + w_z^2) = 0, \tag{2.61a}$$

$$h(w_{\rho\rho} + w_{zz} - \rho^{-1}w_{\rho}) - 2h_{\rho}w_{\rho} - 2h_z w_z = 0. \tag{2.61b}$$

Equations (2.13*a, b*) then transform as follows:

$$\mu_{\rho} = h^{-1}h_{\rho} + \tfrac{1}{2}\rho h^{-2}(h_{\rho}^2 - h_z^2) - \tfrac{1}{2}\rho^{-1}h^{-2}(w_{\rho}^2 - w_z^2), \tag{2.62a}$$

$$\mu_z = h^{-1}h_z + \rho h^{-2}h_{\rho}h_z - \rho^{-1}h^{-2}w_{\rho}w_z. \tag{2.62b}$$

We obtain the approximate solution by the following procedure. We assume the solution we are seeking is $S(\sigma, \zeta)$, depending on the harmonic functions σ and ζ, so that $S(\sigma, 0)$ is the Weyl solution and $S(0, \zeta)$ is the Papapetrou solution. Next we replace σ and ζ by $\lambda\sigma$ and $\lambda\zeta$ where λ is a constant parameter. This is permissible since $\lambda\sigma$ and $\lambda\zeta$ are also harmonic for constant λ. We then expand $S(\lambda\sigma, \lambda\zeta)$ in a power series in λ. This procedure amounts to expanding the functions h and w in a power series in λ and solving (2.61*a, b*) successively in terms of two harmonic functions. Thus we expand h and w as follows:

$$h = 1 + \lambda h^{(1)} + \lambda^2 h^{(2)} + \cdots + \lambda^n h^{(n)} + \cdots, \tag{2.63a}$$

$$w = \lambda w^{(1)} + \lambda^2 w^{(2)} + \cdots + \lambda^n w^{(n)} + \cdots \tag{2.63b}$$

where $h^{(i)}$ and $w^{(i)}$ are functions of ρ and z, and $i = 1, 2, \ldots$. We assume that when $\lambda = 0$ we get Minkowski space as this amounts to setting σ and ζ equal to zero, that is, we assume that $S(0,0)$ gives Minkowski space. This is consistent with the fact that in (2.63a, b) $h = 1$ and $w = 0$ when $\lambda = 0$, for these values of h and w give Minkowski space.

Next we introduce the constant λ in the Weyl and Papapetrou solutions by replacing σ and ζ respectively by $\lambda\sigma$ and $\lambda\zeta$ as follows:

$$h = \exp(\lambda\sigma), \quad w = 0, \tag{2.64a}$$

$$h = \alpha\cosh\lambda\zeta_z + \beta\sinh\lambda\zeta_z, \quad w = (\alpha^2 - \beta^2)^{1/2}\lambda\rho\zeta_\rho. \tag{2.64b}$$

Equations (2.64a, b) give respectively the Weyl and Papapetrou solutions (see (2.19) and (2.34); we have a trivial change of sign in σ). Our object is to obtain a power series solution in terms of harmonic functions σ and ζ which reduces, in each order, to the Weyl solution when $\zeta = 0$ and to the Papapetrou solution when $\sigma = 0$. We set $\alpha = 1$ henceforth in (2.64b), since when $\lambda = 0$, the function h in (2.63a) becomes unity whereas from (2.64b) $h = \alpha$. For the Weyl solution we have

$$h^{(n)} = \frac{1}{n!}\sigma^n, \quad w^{(n)} = 0, \quad \text{all} \quad n. \tag{2.65}$$

and for the Papapetrou solution one gets

$$h^{(1)} = \beta\zeta_z, \quad h^{(2)} = \tfrac{1}{2}\zeta_z^2, \quad h^{(3)} = \tfrac{1}{6}\beta\zeta_z^3, \dots, \tag{2.66a}$$

$$w^{(1)} = (1 - \beta^2)^{1/2}\rho\zeta_\rho, \quad w^{(n)} = 0, \quad n \geqslant 2. \tag{2.66b}$$

The procedure followed here is a somewhat modified form of that followed in Islam (1976a). where the parameter λ was introduced in a different manner and we considered only even powers of λ in h and only odd powers in w. We now substitute the power series (2.63a, b) in (2.61a, b) and equate to zero successive powers of λ. We shall choose the arbitrary constants at each stage so that the series solution reduces to those of Weyl and Papapetrou. To first order we get

$$h_{\rho\rho}^{(1)} + h_{zz}^{(1)} + \rho^{-1}h_\rho^{(1)} = 0, \quad w_{\rho\rho}^{(1)} + w_{zz}^{(1)} - \rho^{-1}w_\rho^{(1)} = 0, \tag{2.67}$$

the solutions to which can be taken as

$$h^{(1)} = \sigma + \beta\zeta_z, \quad w^{(1)} = (1 - \beta^2)^{1/2}\rho\zeta_\rho, \tag{2.68}$$

where σ, ζ (and hence ζ_z) are harmonic, and $|\beta| < 1$. In the second order we get the following equation from (2.63a)

$$\nabla^2 h^{(2)} + h^{(1)}\nabla^2 h^{(1)} - h_\rho^{(1)2} - h_z^{(1)2} - \rho^{-2}(w_\rho^{(1)2} + w_z^{(1)2}) = 0. \tag{2.69}$$

where ∇^2 is defined by (2.18). With the use of the expressions (2.68) for $h^{(1)}$, (2.69) reduces to the following equation

$$\nabla^2 h^{(2)} = \sigma_\rho^2 + \sigma_z^2 + \zeta_{\rho z}^2 + \zeta_{zz}^2 + 2\beta(\sigma_\rho\zeta_{\rho z} + \sigma_z\zeta_{zz}), \tag{2.70}$$

which has the solution

$$h^{(2)} = \tfrac{1}{2}\sigma^2 + \tfrac{1}{2}\zeta_z^2 + \beta\sigma\zeta_z, \tag{2.71}$$

where we have ignored an arbitrary additive harmonic function. This solution is obtained by the application of the identity

$$\nabla^2(FG) = F\nabla^2 G + G\nabla^2 F + (F_\rho G_\rho + F_z G_z), \tag{2.72}$$

for any two functions F and G.

In the second order (2.63b) yields the following equation:

$$\Delta w^{(2)} + h^{(1)}\Delta w^{(1)} - 2h_\rho^{(1)} w_\rho^{(1)} - 2h_z^{(1)} w_z^{(1)} = 0 \tag{2.73}$$

where $\Delta \equiv (\partial^2/\partial\rho^2 + \partial^2/\partial z^2 - \rho^{-1}\partial/\partial\rho)$. After substitution for $h^{(1)}$ and $w^{(1)}$ from (2.68) this reduces to the following equation:

$$\Delta w^{(2)} = 2(1 - \beta^2)^{1/2}\rho(\sigma_z\zeta_{\rho z} - \sigma_\rho\zeta_{zz}). \tag{2.74}$$

The solution of this equation is given by the following two equations through a simple integration:

$$w_\rho^{(2)} = (1 - \beta^2)^{1/2}\rho(\sigma_z\zeta_z - \sigma\zeta_{zz}), \tag{2.75a}$$
$$w_z^{(2)} = (1 - \beta^2)^{1/2}\rho(\sigma\zeta_{\rho z} - \sigma_\rho\zeta_z). \tag{2.75b}$$

The consistency of (2.75a, b) can be shown as follows. From (2.75a) we get, by differentiation with respect to z,

$$w_{\rho z}^{(2)} = (1 - \beta^2)^{1/2}\rho(\sigma_{zz}\zeta_z - \rho\zeta_{zzz}). \tag{2.76}$$

From (2.75b), on differentiation with respect to ρ we get

$$w_{z\rho}^{(2)} = (1 - \beta^2)^{1/2}[\sigma(\zeta_{\rho z} + \sigma\zeta_{\rho\rho z}) - \zeta_z(\sigma_\rho + \rho\sigma_{\rho\rho})]. \tag{2.77}$$

Because σ and ζ are harmonic, we have

$$\zeta_{\rho z} + \rho\zeta_{\rho\rho z} = -\rho\zeta_{zzz}, \quad \sigma_\rho + \rho\sigma_{\rho\rho} = -\rho\sigma_{zz}. \tag{2.78}$$

Substitution from (2.78) and (2.77) shows that (2.76) and (2.77) are the same.

To verify that (2.75a, b) constitute a solution of (2.74), we form the expression $\Delta w^{(2)}$ by adding the ρ-derivative of (2.75a) to the z-derivative of (2.75b) and subtracting from it ρ^{-1} times (2.75a). The result is (2.74).

We will state the results for the third order. The function $h^{(3)}$ can be written as follows:

$$h^{(3)} = h'^{(3)} + \tfrac{1}{6}\sigma^3 + \tfrac{1}{2}\beta\sigma^2\zeta_z + \tfrac{1}{6}\beta\zeta_z^3, \tag{2.79a}$$

where $h'^{(3)}$ satisfies the following equation

$$\nabla^2 h'^{(3)} = (1 + 2\beta^2)\sigma(\zeta_{\rho z}^2 + \zeta_{zz}^2). \tag{2.79b}$$

The function $w^{(3)}$ satisfies the following equation:

$$\Delta w^{(3)} = 2(1 - \beta^2)^{1/2}\rho(\sigma + \beta\zeta_z)(\sigma_z\zeta_{\rho z} - \sigma_\rho\zeta_{zz}). \tag{2.80}$$

It does not seem possible to solve (2.79b) and (2.80) explicitly in terms of harmonic functions. Equation (2.79b) is a Poisson equation, and (2.80) can also be transformed into a Poisson equation with the use of the identity

$$w^{(3)} = \rho w_\rho^{\prime(3)}, \quad \Delta w^{(3)} = \rho(\nabla^2 w^{\prime(3)})_\rho. \qquad (2.81)$$

The Poisson equations can be solved implicitly by the standard integral representation, as in (1.4). At each stage of the approximation one gets a pair of Poisson equations in which the right hand sides (the sources) are given in terms of lower order functions. Thus a power series solution exists in which the terms are given implicitly (in terms of integral representations) by two harmonic functions. However, such a power series is not very useful if one is looking for an exact solution in closed form like, for example, the Weyl or Lewis or Papapetrou solutions. The power series solution does indicate that a general solution depends on at least two harmonic functions. It is easily verified that our series solution reduces to the Weyl and Papapetrou solutions given by (2.65) and (2.66a, b) respectively in the appropriate limits. For simple harmonic functions, such as those given by $\sigma \sim r^{-1}, \zeta \sim r^{-1}, r = (\rho^2 + z^2)^{1/2}$, the above scheme can be carried out explicitly to arbitrarily high orders.

To consider physical interpretation, let the harmonic functions σ and ζ be chosen as follows (with $r^2 = \rho^2 + z^2$):

$$\sigma = \frac{a'}{r} + \frac{a''z}{r^3} + \cdots, \quad \zeta = \frac{b'}{r} + \frac{b''z}{r^3} + \cdots, \qquad (2.82)$$

where a', a'', b', b'' are constants and dots represent harmonic functions which vanish at infinity faster than the terms preceding the dots. Then the approximate solution reads as follows, to second order:

$$h = 1 + \lambda \left[\frac{a'}{r} + \frac{(a'' - b'\beta)z}{r^3} + \cdots \right] + \lambda^2 \left(\frac{a'^2}{2r^2} + \cdots \right), \qquad (2.83a)$$

$$w = \lambda(1 - \beta^2)^{1/2} \left(-\frac{b'\rho^2}{r^3} - \frac{3b''\rho^2 z}{r^5} + \cdots \right)$$
$$+ \lambda^2 (1 - \beta^2)^{1/2} \left(-\frac{a'b'\rho^2}{r^4} + \cdots \right), \qquad (2.83b)$$

where we have ignored terms which vanish at infinity faster than r^{-2}. As can be seen by comparing with (2.36) (w and k have the same leading behaviour) the functions h and w display the correct boundary conditions at infinity for a bounded source. Clearly the constants a', b' are related to the mass and angular momentum respectively of the source. The constants a'', b'' and the harmonic functions which are represented by the dots describe the detailed structure of the source. These terms represent the higher 'multiple moments' of the mass and angular momentum. The solution could thus describe the approximate field of quite a general bounded source.

The harmonic function ζ represents the effect of rotation in some sense because the metric becomes static (Weyl) when $\zeta = 0$. If it is required that the field should tend to spherical symmetry in the limit $\zeta \to 0$, this can be achieved by taking σ to be the harmonic function which yields the Schwarzschild solution (i.e., (2.21) with $\delta = 1$) in the limit $\zeta \to 0$. This can be done in a variety of ways. Recall that the harmonic function σ leading to the Schwarzschild solution is the Newtonian potential of a uniform rod of a certain 'unit' density. Let the constants $a_0, a_1, \ldots, a_N$ (N could be infinite) parametrize the harmonic function ζ such that $\zeta \to 0$ as $a_n \to 0$; for example, the a_n could be the constants in the following expression for ζ:

$$\zeta = \sum_{n=0}^{N} a_n \left(\frac{\partial}{\partial z} \right)^n \frac{1}{r}, \quad r^2 = \rho^2 + z^2. \tag{2.84}$$

Now consider a (Newtonian) bounded axisymmetric distribution of matter given by these parameters in such a way that as the a_n tend to zero this distribution of matter (call this configuration Σ) tends to an infinitely thin uniform rod of unit density. The constants could represent either variable density or the shape of the configuration Σ. Now let $-\sigma$ be the exterior Newtonian potential of Σ. Then such a choice for σ would be a possible one if it is required that the field should tend to spherical symmetry as the $a_n \to 0$, that is, as $\zeta \to 0$. For example, $-\sigma$ could be the Newtonian potential of a finite rod with variable density such that the density tends to unity as $\zeta \to 0$, or it could be the potential of a cylinder which tends to a uniform (thin) rod as $\zeta \to 0$. (We take $-\sigma$ because in our approximate solution we have used $h = f^{-1}$ rather than f.) Thus our approximate solution could represent quite general physical requirements.

The exterior gravitational field of an axisymmetric Newtonian distribution of matter is given by a single axisymmetric harmonic function, whether or not the matter is rotating. Why is it necessary to have at least two harmonic functions to describe the exterior of a rotating source in general relativity as seems to be implied by the above analysis? The partial answer is connected with remarks made earlier in connection with Fig. 1.2. Unlike in Newtonian gravitation, in general relativity any kind of motion of matter produces extra gravitational fields analogous to the magnetic field produced by moving charges in electromagnetism. It is interesting that a harmonic function makes its appearance as in Newtonian gravitation. We shall see the pervasive influence of harmonic functions also later on in the book.

For a realistic field one must have $w \to 0$ (i.e., no rotation) as the mass of the source tends to zero. This is not true, for example, of the Papapetrou solutions, but can be achieved in the approximate solution by, for example, choosing the constants b', β'', etc. in ζ to be proportional to a', the coefficient of the leading term in σ (see (2.82)). For general (harmonic) σ and ζ the approximate solutions could represent, to this order, rotating Weyl fields which would include, for example, rotating axisymmetric but non-spherical rigid bodies. The constant λ represents the strength of the gravitational field in the sense that when $\lambda = 0$ we have Minkowski space with no gravitational fields.

After this excursion into approximate solutions, we shall consider some important asymptotically flat exact rotating solutions of (2.12a, b) in the next chapter.

3

The Kerr and Tomimatsu–Sato solutions

3.1. Introduction

One of the simplest situations for a bounded rotating source in Newtonian theory of gravitation is the case of a homogenous inviscid fluid mass rotating uniformly which was considered briefly in Section 1.1. In this case both the interior and exterior Newtonian gravitational potentials are known explicitly and this case has been studied extensively (see, for example, Chandrasekhar, 1969). In general relativity the finding of an exact solution of Einstein's equations which represents a uniformly rotating homogeneous inviscid fluid mass – either the interior or the exterior field – presents formidable problems and we are far from finding such an exact solution, if one exists. Some progress has been made for finding an approximate solution for this case (see, for example, Chandrasekhar, 1971, Bardeen, 1971). In general, finding exact solutions of Einstein's equations for well-defined physical situations is extremely difficult and very few such solutions are known. The gravitational field of a uniformly rotating bounded source must depend on at least two variables. Finding *any* solutions of Einstein's equations depending on two or more variables is quite difficult, let alone a physically interesting one. The first exact solution of Einstein's equations to be found which could represent the exterior field of a bounded rotating source was that of Kerr (1963). An essential property of such a solution is that it should be asymptotically flat, since the gravitational field tends to zero as one moves further and further away from the source. The Kerr solution was the first known rotating solution which was asymptotically flat with a source which has non-zero mass. A class of zero-mass asymptotically flat rotating solutions described in the last chapter was found by Papapetrou earlier (1953). However, no interior solution has yet been found which matches smoothly onto the Kerr solution. It is believed that the Kerr solution represents the exterior gravitational field of a highly collapsed rotating star – a rotating black hole.

Although many stationary axially symmetric exact solutions of Einstein's equations (equations $(2.12a, b)$) are known, very few of these are asymptotically flat, and so their physical interpretation is uncertain. The first rotating asymptotically flat solutions to be found after the Kerr solution were the Tomimatsu–Sato solutions (1972, 1973). These solutions differ from the Kerr solution in one important respect. The Kerr solution has the property that when the angular momentum of the source producing the field tends to zero, the solution tends to the Schwarzschild solution, representing the exterior field of a spherically symmetric source. This behaviour is what one would expect for realistic stars, because for the latter the departure from spherical symmetry is usually caused by rotation and if the rotation vanishes one would get a spherically symmetric star, whose exterior field is the Schwarzschild solution. However, the Tomimatsu–Sato solutions do not tend to the Schwarzschild solution when the angular momentum parameter of the source tends to zero. Instead, these solutions tend to the non-spherically symmetric static Weyl solutions given by (2.18)–(2.21) with $\delta = 2, 3 \ldots$. Thus the Tomimatsu–Sato solutions are not so suitable as exterior fields of realistic astrophysical bodies. Nevertheless, their asymptotic flatness makes them very interesting exact solutions.

There is no easy way to derive the Kerr solution (see, for example, Chandrasekhar (1978)). Here we shall derive the Kerr solution from Ernst's (1968a) form of $(2.12a, b)$, since the Tomimatsu–Sato (TS) solutions can also be derived by an extension of this method.

3.2. Ernst's form of Einstein's equations for rotating fields

Equation $(2.12b)$ can be written as follows:

$$(\rho^{-1} f^2 w_\rho)_\rho + (\rho^{-1} f^2 w_z)_z = 0, \tag{3.1}$$

which implies the existence of a function u such that (this u is distinct to that used in Section 2.6)

$$u_\rho = - \rho^{-1} f^2 w_z, \quad u_z = \rho^{-1} f^2 w_\rho. \tag{3.2}$$

In terms of f and u $(2.12a)$ can be written as follows:

$$f \nabla^2 f - f_\rho^2 - f_z^2 + u_\rho^2 + u_z = 0, \tag{3.3a}$$

while from (3.2) we get, by eliminating w, the following equation:

$$f \nabla^2 u = 2 f_\rho u_\rho + 2 f_z u_z. \tag{3.3b}$$

In terms of f and u $(2.13a, b)$ can be written as follows:

$$\mu_\rho' = \tfrac{1}{2} \rho f^{-2}(f_\rho^2 - f_z^2) + \tfrac{1}{2} \rho f^{-2}(u_\rho^2 - u_z^2), \tag{3.4a}$$

$$\mu_z' = \rho f^{-2} f_\rho f_z + \rho f^{-2} u_\rho u_z. \tag{3.4b}$$

where $\mu' = \mu + \log f$. Now define the complex function E as follows:

$$E = f + iu. \tag{3.5}$$

It can be verified that (3.3a, b) are the real and imaginary parts of the single complex equation

$$(\text{Re}E)\nabla^2 E = E_\rho^2 + E_z^2, \tag{3.6}$$

where ∇^2 is the Laplacian operator defined by (2.18) and Re E = real part of $E = f$. Next introduce a new unknown function ξ instead of E defined by

$$E = \frac{\xi - 1}{\xi + 1}. \tag{3.7}$$

Then (3.6) becomes

$$(\xi\xi^* - 1)\nabla^2\xi = 2\xi^*(\xi_\rho^2 + \xi_z^2), \tag{3.8}$$

where ξ^* is the complex conjugate of ξ. Instead of the variables (ρ, z) we introduce prolate spheroidal coordinates (x, y) as follows:

$$\rho = (x^2 - 1)^{1/2}(1 - y^2)^{1/2}, \quad z = xy, \tag{3.9}$$

which can be solved for x, y as follows:

$$x = \tfrac{1}{2}(R^{(+)} + R^{(-)}), \quad y = \tfrac{1}{2}(R^{(+)} - R^{(-)}), \tag{3.10a}$$

$$R^{(\pm)} = [\rho^2 + (z \pm 1)^2]^{1/2}. \tag{3.10b}$$

In terms of x and y, (3.8) can be written as follows:

$$(\xi\xi^* - 1)[(x^2 - 1)\xi_{xx} + 2x\xi_x + (1 - y^2)\xi_{yy} - 2y\xi_y]$$
$$= 2\xi^*[(x^2 - 1)\xi_x^2 + (1 - y^2)\xi_y^2]. \tag{3.11}$$

Equations (3.8) and (3.11) are the Ernst form of Einstein's equations (2.12a, b). In the variables x and y (3.2) and (3.4a, b) can be written as follows:

$$w_x = (1 - y^2)f^{-2}u_y, \quad w_y = (1 - x^2)f^{-2}u_x, \tag{3.12}$$

$$\mu_x' = \frac{(1 - y^2)f^{-2}}{2(x^2 - y^2)}[x(x^2 - 1)(f_x^2 + u_x^2) + x(y^2 - 1)(f_y^2 + u_y^2)$$
$$- 2y(x^2 - 1)(f_x f_y + u_x u_y)], \tag{3.13a}$$

$$\mu_y' = \frac{(x^2 - 1)f^{-2}}{2(x^2 - y^2)}[y(x^2 - 1)(f_x^2 + u_x^2) - y(1 - y^2)(f_y^2 + u_y^2)$$
$$+ 2x(1 - y^2)(f_x f_y + u_x u_y)]. \tag{3.13b}$$

We leave the derivation of (3.11)–(3.13b) as an exercise for the reader.

3.3. The Kerr solution

A striking feature of the Ernst equation (3.11) is that the Kerr solution is given by the following simple solution of it:

$$\xi = px - iqy \tag{3.14}$$

where p and q are constants with $p^2 + q^2 = 1$. From (3.5), (3.7) and (3.14) one sees that for this solution

$$f = \frac{p^2x^2 + q^2y^2 - 1}{(px + 1)^2 + q^2y^2}, \quad u = \frac{-2qy}{(px + 1)^2 + q^2y^2}. \tag{3.15}$$

From (3.12) and (3.15) we get w as follows:

$$w = \frac{2qp^{-1}(1 - y^2)(px + 1)}{p^2x^2 + q^2y^2 - 1} + w_0 \tag{3.16}$$

where w_0 is a constant. Equations (3.13a, b) yield μ' as follows:

$$e^{\mu'} = \frac{A(x^2p^2 + q^2y^2 - 1)}{x^2 - y^2}, \tag{3.17}$$

where A is an arbitrary constant. From (3.17) and the definition of μ' $(e^{\mu'} = f e^\mu)$ we get

$$e^\mu = \frac{A[(px + 1)^2 + q^2y^2]}{x^2 - y^2}. \tag{3.18}$$

We now introduce coordinates r and θ related to x and y as follows (the r, θ here are not necessarily the same as those used in the last chapter):

$$px + 1 = pr, \quad y = \cos\theta. \tag{3.19}$$

We also introduce constants m and a related to p and q as follows:

$$p^{-1} = m, \quad p^{-1}q = a, \quad m^2 - a^2 = 1. \tag{3.20}$$

The mass and angular momentum of the Kerr solution will turn out to be m and ma respectively, these constants being evaluated here in units such that m and a are related as in (3.20). To transform the Kerr solution to its standard form, that of Boyer and Lindquist (1967), we will start with the form (1.96), which we will write again for convenience

$$ds^2 = f\,dt^2 - 2k\,d\phi\,dt - l\,d\phi^2 - e^\mu(d\rho^2 + dz^2). \tag{1.96}$$

For the Kerr solution $f, k(= fw), e^\mu$ are given by (3.15), (3.16) and (3.18) but these functions are given in terms of x and y. The function l is $f^{-1}(\rho^2 - k^2)$ (see (2.9)). With the use of (3.9), (3.19) and (3.20), we can express (ρ, z) in terms of (r, θ) defined by (3.19), as follows:

$$\rho = (r^2 - 2mr + a^2)^{1/2}\sin\theta, \quad z = (r - m)\cos\theta. \tag{3.21}$$

With the use of (3.19)–(3.21) e^μ given by (3.18) can be written as follows:

$$e^\mu = \frac{A}{m^2}\frac{r^2 + a^2\cos^2\theta}{(r - m)^2 + (a^2 - m^2)\cos^2\theta}. \tag{3.22}$$

With the use of (2.9), (3.15), (3.16), (3.21) we get

$$d\rho^2 + dz^2 = [(r - m)^2 + (a^2 - m^2)\cos^2\theta][(r^2 - 2mr + a^2)^{-1}dr^2 + d\theta^2],$$
$$(3.23)$$

$$k = 2amr\sin^2\theta(r^2 + a^2\cos^2\theta)^{-1},$$
$$(3.24)$$

$$l = 2ma^2r\sin^4\theta(r^2 + a^2\cos^2\theta)^{-1} + (r^2 + a^2)\sin^2\theta.$$
$$(3.25)$$

In the last line use has been made of the following identity:

$$(r^2 - 2mr + a^2)(r^2 + a^2\cos^2\theta)^2 - 4a^2m^2r^2\sin^2\theta$$
$$= (r^2 - 2mr + a^2\cos^2\theta)(r^4 + r^2a^2 + r^2a^2\cos^2\theta$$
$$- 2ma^2r\cos^2\theta + 2ma^2r + a^4\cos^2\theta).$$
$$(3.26)$$

Combining (3.22)–(3.25), we get the following standard form of the Kerr solution:

$$ds^2 = (1 - 2mr\Sigma_1^{-1})dt^2 - 4amr\sin^2\theta\Sigma_1^{-1}d\phi\,dt$$
$$- (2ma^2r\sin^2\theta\Sigma_1^{-1} + r^2 + a^2)\sin^2\theta\,d\phi^2 - \Sigma_1(\Sigma_2^{-1}dr^2 + d\phi^2)$$
$$(3.27)$$

where $\Sigma_1 \equiv r^2 + a^2\cos^2\theta$, $\Sigma_2 \equiv (r^2 - 2mr + a^2)$. In (3.27) the constants m and a are no longer related by (3.20) as this relation has been used, for example, in (3.22) to convert to general units for m and a. We have set $A = m^2$, and $w_0 = 0$.

For large r the metric (3.27) has the following form:

$$ds^2 \simeq \left(1 - \frac{2m}{r}\right)dt^2 - \left(1 - \frac{2m}{r}\right)^{-1}dr^2 - r^2(d\theta^2 + \sin^2\theta\,d\phi^2)$$
$$- \frac{4am\sin^2\theta}{r}d\phi\,dt.$$
$$(3.28)$$

Changing to coordinates $\rho = r\sin\theta$, $z = r\cos\theta$ and comparing with (1.88) we see that m is the mass and ma is the angular momentum of the source. (The different radial coordinates we have been using coincide at infinity; the transformation used here is the asymptotic form of (3.21).) For $a = 0$, that is, when there is no angular momentum so that the source is non-rotating, the metric (3.27) reduces to the Schwarzschild metric (2.22).

Writing $(x^0, x^1, x^2, x^3) = (t, r, \theta, \phi)$, the inverse of the metric (3.27) can be written as follows:

$$\left.\begin{array}{l} g^{00} = (\Sigma_1\Sigma_2)^{-1}[(r^2 + a^2)^2 - \Sigma_2a^2\sin^2\theta], \\[4pt] g^{11} = -\Sigma_1^{-1}\Sigma_2, \quad g^{22} = -\Sigma_1^{-1}, \\[4pt] g^{33} = -(\Sigma_1\Sigma_2\sin^2\theta)^{-1}(\Sigma_1 - 2mr), \quad g^{03} = (\Sigma_1\Sigma_2)^{-1}(2mar). \end{array}\right\}$$
$$(3.29)$$

For the following discussion we assume $m^2 > a^2$. In the Schwarzschild metric the horizon is determined by the equation $g_{00} = 0$. In the Kerr solution this equation corresponds to the surfaces

$$r^2 + a^2 \cos^2 \theta - 2mr = 0 \qquad (3.30)$$

or

$$r = m \pm (m^2 - a^2 \cos^2 \theta)^{1/2}. \qquad (3.30a)$$

The surface $g_{00} = 0$ in the Schwarzschild metric is a null surface, which means that it is a surface of the form $F = 0$ with

$$g^{\mu\nu} F_{,\mu} F_{,\nu} = 0. \qquad (3.31)$$

It can be verified from (3.29), (3.30) and (3.31) that (3.30) is not a null surface. Thus (3.30) is not the horizon for the Kerr metric. Consider instead the surfaces $\Sigma_2 = 0$, that is,

$$r = m \pm (m^2 - a^2)^{1/2}. \qquad (3.32)$$

Call these surfaces Σ_+ and Σ_-. It can be verified that the surfaces (3.32) indeed satisfy (3.31) so that they are null surfaces. No outgoing null or time-like geodesics cross the surface Σ_+, so that Σ_+ is the horizon for the Kerr metric.

Let us call the surfaces (3.30a) S_+ and S_-. The meaning of these is as follows. The Killing vector corresponding to the time independence of (3.27) is $\partial/\partial t$, that is, it is the vector $\xi^\mu = (1, 0, 0, 0)$. This vector is time-like only outside S_+ or inside S_-, it is null on $S_\pm$ and space-like in between S_+ and S_-. The surface S_+ is called the stationary limit surface since it is only outside this surface that a material particle (such as a spaceship) can remain at rest with respect to infinity. The surface S_+ is time-like except at two points on the axis when it coincides with Σ_+ and where it is null. The region between

Fig. 3.1. Illustration of the central regions of the Kerr metric.

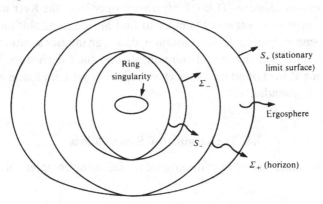

S_+ and Σ_+ is called the ergosphere. Particles can escape to infinity from this region but not from inside Σ_+. Also in the ergosphere it is impossible for a material particle or light wave to remain at rest with respect to observers at infinity. The metric (3.27) also has a 'ring' singularity within the surface S_-. These properties are summarized in Fig. 3.1. The surfaces $S_\pm$ and $\Sigma_\pm$ are all non-singular. Inside the surface Σ^- one gets closed time-like curves so one gets violation of causality and thus unphysical behaviour. Such violation of causality does not occur outside Σ^-. Thus the unphysical region is 'covered' by the region between Σ^- and Σ^+, from which material particles and signals cannot emerge to the region outside Σ^+, to communicate with a distant observer. For this reason the unphysical nature of the geometry within Σ^- is thought to be acceptable, and the Kerr solution for $m^2 > a^2$ is believed to represent the field of a highly collapsed rotating star – a rotating black hole. For $a^2 > m^2$ violations of causality occur in regions accessible to distant observers and hence in this case the metric is unphysical.

When $m = 0$ the central gravitating mass is absent and so the metric should reduce to the Minkowski metric. In fact (3.27) reduces to

$$ds^2 = dt^2 - (r^2 + a^2 \cos^2 \theta)[(r^2 + a^2)^{-1} dr^2 + d\theta^2]$$
$$ - (r^2 + a^2) \sin^2 \theta \, d\phi^2. \qquad (3.33)$$

It is readily seen that (3.33) reduces to the Minkowski metric in cartesian coordinates (x, y, z) given by

$$x = (r^2 + a^2)^{1/2} \sin \theta \cos \phi, \qquad (3.34a)$$
$$y = (r^2 + a^2)^{1/2} \sin \theta \sin \phi, \qquad (3.34b)$$
$$z = r \cos \theta. \qquad (3.34c)$$

We refer to the books by Hawking and Ellis (1973), Landau and Lifshitz (1975), Bose (1980), Chandrasekhar (1983) and also to Carter (1968, 1972) and Boyer and Lindquist (1967) for further properties of the Kerr solution.

There have been various attempts to find an interior solution which matches smoothly onto the Kerr solution, that is, an interior solution with a boundary at which the metric components and their derivatives will be continuous at the boundary. No such solution has yet been found (see, for example, Hernandez, 1967).

3.4. The Tomimatsu–Sato solutions

In 1972 Tomimatsu and Sato discovered the following complex solution of (3.11):

$$\xi = \frac{p^2 x^4 + q^2 y^4 - 2ipqxy(x^2 - y^2) - 1}{2px(x^2 - 1) - 2iqy(1 - y^2)},\qquad(3.35)$$

where p and q are constants with $p^2 + q^2 = 1$. From (3.5) and (3.7) we see that if we write ξ in the following form (see the Appendix):

$$\xi = \frac{N + iN'}{D + iD'},\qquad(3.36)$$

where N, N', D, D' are real polynomials in x, y, then

$$f = \frac{N^2 + N'^2 - D^2 - D'^2}{(N + D)^2 + (N' + D')^2},\qquad(3.37a)$$

$$u = \frac{2(N'D - ND')}{(N + D)^2 + (N' + D')^2},\qquad(3.37b)$$

We can then substitute in (3.12), (3.13a, b) from (3.37a, b) and obtain the corresponding w and μ, noting that for (3.35) we have

$$N \equiv p^2 x^4 + q^2 y^4 - 1, \quad N' \equiv -2pqxy(x^2 - y^2),\qquad(3.38a)$$

$$D \equiv 2px(x^2 - 1), \quad D' = -2qy(1 - y^2).\qquad(3.38b)$$

The functions f, w, μ in the metric

$$ds^2 = f(dt - w\,d\phi)^2 - \rho^2 f^{-1}\,d\phi^2 - e^\mu(d\rho^2 + dz^2),\qquad(3.39)$$

can be written as

$$f = \frac{A}{B}, \quad w = \frac{2p^{-1}q(1 - y^2)C}{A}, \quad e^\mu = \frac{B}{p^4(x^2 - y^2)^4},\qquad(3.40)$$

where

$$A \equiv [p^2(x^2 - 1)^2 + q^2(1 - y^2)^2]^2 - 4p^2 q^2(x^2 - 1)(1 - y^2)(x^2 - y^2)^2,\qquad(3.41a)$$

$$B \equiv (p^2 x^4 + q^2 y^4 - 1 + 2px^3 - 2px)^2 + 4q^2 y^2(px^3 - pxy^2 + 1 - y^2)^2,\qquad(3.41b)$$

$$C \equiv p^2(x^2 - 1)[(x^2 - 1)(1 - y^2) - 4x^2(x^2 - y^2)]$$
$$- p^3 x(x^2 - 1)[2(x^4 - 1) + (x^2 + 3)(1 - y^2)] + q^2(1 + px)(1 - y^2)^2.\qquad(3.41c)$$

When $q = 0$, we get $w = 0$, so there is no rotation and ξ reduces (with $p = 1$) to

$$\xi = \frac{x^2 + 1}{2x}.\qquad(3.42)$$

With the use of (3.7) and (3.10a, b) we see that (3.42) implies

$$f = \frac{(x - 1)^2}{(x + 1)^2} = \left(\frac{R^{(+)} + R^{(-)} - 2}{R^{(+)} + R^{(-)} + 2}\right)^2\qquad(3.43)$$

which gives the Weyl solution (2.21) with $\delta = 2$ and $m = 1$ (the mass of the solution in this case is $2m = 2$ units). Thus in the absence of rotation the Tomimatsu–Sato (TS) solution given by (3.35) reduces to the axially symmetric static Weyl solution given by (2.21) with $\delta = 2$. (Tomimatsu and Sato prefer to use ρ, z coordinates which are proportional to those given by (3.9).)

From (3.10a, b) and (3.19) we see that for large values of x the coordinate r and the expression $(\rho^2 + z^2)^{1/2}$ and x are asymptotically all equal. Using this fact and that y is $\cos \theta$ it is readily verified that f and w given by (3.40) have the correct asymptotic behaviour at infinity given by (1.88). For example, f can be written as

$$f = \frac{p^4 x^8 + \cdots}{p^4 x^8 + 4p^3 x^7 + \cdots} = 1 - \frac{4}{px} + O(x^{-2}), \qquad (3.44)$$

where dots in (3.44) represent terms in x of degree 6 and less. Thus the mass M of the solution (3.40) is given by $2p^{-1}$, which is consistent with the fact that when $q \to 0$, we get $p \to 1$ and obtain the Weyl solution (3.43) with two units of mass. Equation (3.40) also implies

$$w = \frac{2p^{-1}q(1 - y^2)(- 2p^3 x^7 + \cdots)}{p^4 x^8 + 4p^3 x^7 + \cdots} = - \frac{4q \sin^2 \theta}{p^2 x} + O(x^{-2}), \qquad (3.45)$$

where again dots represent terms in x of degree six and less. Again comparing with (1.88) we see that the angular momentum J of the solution is given by

$$J = \frac{4q}{p^2} = qM^2. \qquad (3.46)$$

The solution given by (3.40) is thus asymptotically flat and for this reason is of considerable interest.

In 1973 Tomimatsu and Sato found two other solutions which in the absence of rotation reduce to the solution of (3.11) given by

$$\xi = \frac{(x + 1)^\delta + (x - 1)^\delta}{(x + 1)^\delta - (x - 1)^\delta}, \qquad (3.47)$$

given by $\delta = 3, 4$. Equation (3.47) with $\delta = 3, 4$ can be seen to yield the Weyl solutions (2.21) with $\delta = 3, 4$. The corresponding full solutions for ξ and f, w, μ are very complicated and are given explicitly in Tomimatsu and Sato (1973). The constant δ is referred to as the deformation parameter because the value of δ represents the departure from spherical symmetry of the static limit of these solutions since $\delta = 1$ represents the Schwarzschild solution.

Yamazaki (1977a) made a conjecture as to the analytic form of the TS solution for the general integral δ, that is, the TS solution that reduces in the absence of rotation to the Weyl solution (2.21) with integral δ. We can call this solution TS(δ). Thus TS(1) is the Kerr solution, and Tomimatsu and Sato (1972, 1973) found TS(2), TS(3) and TS(4). Yamazaki made a conjecture as to the analytic form for TS(δ) for integral δ and Hori (1978) was able to prove this conjecture to be correct.

Several people have studied properties of the TS solutions. Gibbons and Russell-Clarke (1973) showed that TS(2) has a 'ring' singularity on the equatorial plane $y = 0$ given by

$$B(x, 0) = 0, \quad y = 0, \tag{3.48}$$

where B is given by (3.41b). They further showed that this metric possesses 'naked' singularities, that is, singularities which are not 'hidden' by event horizons and are accessible to distant observers, and that the metric TS(2) is of a general Petrov type (see, for example, Kramer *et al.* (1980), for a definition of Petrov types). Kinnersley and Kelley (1974) studied some limits of TS metrics including the limit $q = 1$. They were in this way led to a two-parameter family of exact solutions which, however, are not asymptotically flat. Ernst (1976) found a new representation for the TS solutions and with its use studied some properties of geodesics. Geodesics were also considered by Tomimatsu and Sato (1973). Economou (1976) studied the TS(2) solution in the neighbourhood of the 'poles' $x = 1, y = \pm 1$. Yamazaki (1977b) studied 'ring' singularities of TS(δ) for an arbitrary integral δ. Hoenselaers and Ernst (1983) studied geodesics in the TS metrics and showed that none of the non-equatorial geodesics reach the ring singularities on the equatorial plane.

As to other metrics generalizing the TS metrics, Kinnersley and Chitre (1978a, b) generalized the TS(2) solution to a five-parameter family of solutions. From this five-parameter family they obtained a two-parameter family of metrics which is an asymptotically flat rotating solution. Cosgrove (1977) generalized the TS(δ) solutions to a three-parameter family of solutions in which one of the parameters is δ. However, for non-integral δ the solution is not given in closed form like that of Yamazaki–Hori but is given in terms of a solution of two second-order ordinary differential equations.

3.5. Uniqueness of the Kerr solution

Israel (1967) showed that under quite general conditions the only asymptotically flat static (non-rotating) vacuum solution of Einstein's equations that has a non-singular horizon is the Schwarzschild solution. Hawking (1972) showed that the geometry of a stationary space–time with a regular event horizon must be axisymmetric in the exterior of the horizon and the topology of the event horizon must be $S^2 \times R^1$ where S^2 is the surface of a two-sphere and R^1 is the real line. Carter (1971, 1972) showed that in the latter case, under certain conditions (such as asymptotic flatness and a regular event horizon) there exist at most certain families of solutions each depending on at most two parameters, one such family being the Kerr family with $a^2 < m^2$. Robinson (1975) showed that there is indeed a unique family, namely the Kerr family, which is asymptotically flat, stationary, and has a regular event horizon. The significance of this result is that a rotating black hole is described uniquely by the Kerr solution. We are here confining ourselves to the pure Einstein equations with no electromagnetic field. The TS solutions do not violate this theorem for they have 'naked' singularities which are not enclosed by event horizons. We have not stated here the precise conditions of these theorems; these conditions are somewhat technical and we refer to the papers cited for details of them.

4

Rotating neutral dust

4.1. Introduction

A global solution of Einstein's equations is one in which an interior solution is matched smoothly onto an exterior solution, that is the metric and its derivatives are continuous at the boundary. Such global solutions can give considerable insight into the physical content of Einstein's equations and hence of general relativity. However, very few such global solutions are known partly because of the extreme scarcity of physically realistic interior solutions. One of the few global solutions known in general relativity is that of Van Stockum. Referring to this work Bonnor (1980a), who extended Van Stockum's results, says 'In a fine paper, well ahead of its time, Van Stockum (1937) completely solved the problem of a rigidly rotating infinitely long cylinder of dust, including the application of adequate boundary conditions.' In this chapter we shall consider in some detail the problem which Van Stockum solved.

In the first part of his paper Van Stockum found a class of exact interior solutions for rigidly notating neutral (uncharged) dust. By 'dust' here is meant pressureless matter. These solutions are axially symmetric but not necessarily cylindrically symmetric. They are quite distinct from the Van Stockum exterior rotating solutions discussed in Section 2.6. Like the exterior solutions the interior solutions are given in terms of a harmonic function. In the second part of his paper Van Stockum specialized his solution to cylindrical symmetry so that it described an infinitely long cylinder of rigidly rotating dust. For the exterior solution he took the cylindrically symmetric form of the Lewis solutions and his exterior solution discussed in Section 2.6; he then matched these solutions smoothly at the boundary of the cylinder, and thus obtained a global solution. Before discussing his global solution we shall consider Van Stockum's axisymmetric interior solution for rigidly rotating dust.

4.2. Einstein interior equations for rotating dust

As mentioned in Chapter 1, the energy momentum tensor for a perfect fluid is given as follows:

$$T^{\mu\nu} = (\varepsilon + p)u^{\mu}u^{\nu} - pg^{\mu\nu}, \qquad (1.36)$$

where u^{μ} is the four velocity, ε the density of mass energy and p is the pressure. When the pressure is zero there is no energy of the matter due to the random motion of the particles and so the mass-energy density consists of only the density of the particles which is mn, where m is the mass of the particles and n is the number density of the particles. Thus in this case

$$T^{\mu\nu} = mnu^{\mu}u^{\nu}. \qquad (4.1)$$

Einstein's equations (1.35) are given by

$$R^{\mu\nu} - \tfrac{1}{2}g^{\mu\nu}R = 8\pi mnu^{\mu}u^{\nu}. \qquad (4.2)$$

Because the divergence of the left hand side vanishes, we get

$$(nu^{\mu}u^{\nu})_{;\nu} = (nu^{\nu})_{;\nu}u^{\mu} + nu^{\mu}_{;\nu}u^{\nu} = 0. \qquad (4.3)$$

Now $(nu^{\nu})_{;\nu} = 0$ is just the condition of conservation of matter so that we get (with $u^{\mu} = dx^{\mu}/ds$):

$$u^{\mu}_{;\nu}u^{\nu} = \frac{d^2x^{\mu}}{ds^2} + \Gamma^{\mu}_{\nu\sigma}\frac{dx^{\sigma}}{ds}\frac{dx^{\nu}}{ds} = 0, \qquad (4.4)$$

which is the geodesic equation so that particles of the dust follow geodesics.

We again specialize to the general rotating metric given by

$$ds^2 = f\,dt^2 - 2k\,d\phi\,dt - l\,d\phi^2 - e^{\mu}(d\rho^2 + dz^2), \qquad (4.5)$$

where f, k, l and μ are all functions of ρ and z. Setting $(x^0, x^1, x^2, x^3) = (t, \rho, z, \phi)$, the components of the four-velocity of the rotating dust are

$$u^0 = \frac{dt}{ds} = (f - 2\Omega k - \Omega^2 l)^{-1/2}, \quad u^1 = \frac{d\rho}{ds} = 0, \quad u^2 = \frac{dz}{ds} = 0, \\ u^3 = \frac{d\phi}{ds} = \frac{d\phi}{dt}\frac{dt}{ds} = \Omega u^0. \qquad (4.6)$$

We envisage the dust to be rotating about the z-axis (the axis of symmetry) with angular velocity Ω which is in general a function of ρ and z. The dust particles occupy positions given by constant values of ρ and z (hence $u^1 = u^2 = 0$) while only their ϕ coordinate changes ($d\phi/dt = \Omega$).

It is more convenient to consider the following form of (4.2):

$$R_{\mu\nu} = 8\pi mn(u_{\mu}u_{\nu} - \tfrac{1}{2}g_{\mu\nu}). \qquad (4.7)$$

In the metric (4.5) three of the field equations can be written as follows:

$$2e^{\mu}D^{-1}R_{00} = (D^{-1}f_{\rho})_{\rho} + (D^{-1}f_{z})_{z} + D^{-3}f\Sigma$$
$$= 8\pi mnD^{-1}e^{\mu}(f - 2\Omega k - \Omega^{2}l)^{-1}$$
$$\cdot[2\Omega k(-f + \Omega k) + f(f + \Omega^{2}l)], \qquad (4.8a)$$
$$-2e^{\mu}D^{-1}R_{03} = (D^{-1}k_{\rho})_{\rho} + (D^{-1}k_{z})_{z} + D^{-3}k\Sigma$$
$$= 8\pi mnD^{-1}e^{\mu}(f - 2\Omega k - \Omega^{2}l)^{-1}(fk + 2\Omega fl - \Omega^{2}kl), \quad (4.8b)$$
$$-2e^{\mu}D^{-1}R_{33} = (D^{-1}l_{\rho})_{\rho} + (D^{-1}l_{z})_{z} + D^{-3}l\Sigma$$
$$= -8\pi mnD^{-1}e^{\mu}(f - 2\Omega k - \Omega^{2}l)^{-1}(fl + 2k^{2} + 2\Omega kl + \Omega^{2}l^{2})$$
$$(4.8c)$$

where $D^{2} = fl + k^{2}$ and $\Sigma = f_{\rho}l_{\rho} + f_{z}l_{z} + k_{\rho}^{2} + k_{z}^{2}$. An important combination of (4.8a,b,c) is the following:

$$e^{\mu}D^{-1}(lR_{00} - 2kR_{03} - fR_{33}) = D_{\rho\rho} + D_{zz} = 0. \qquad (4.9)$$

Thus D satisfies the same equation as before (see (2.3)). We can therefore follow the same procedure as we did earlier to derive the relation (2.9) between f, k and l:

$$fl + k^{2} = \rho^{2}. \qquad (4.10)$$

The geodesic equations (4.4) imply the following two equations:

$$f_{\rho} - 2\Omega k_{\rho} - \Omega^{2}l_{\rho} = 0, \qquad (4.11a)$$
$$f_{z} - 2\Omega k_{z} - \Omega^{2}l_{z} = 0. \qquad (4.11b)$$

The equations so far are valid for Ω as a function of ρ and z so that we can have differential rotation. We now choose Ω to be constant so that the rotation is rigid. We also transform to functions F, K, L given as follows:

$$L = l, \quad K = k + \Omega l, \quad F = f - 2\Omega k - \Omega^{2}l. \qquad (4.12)$$

This amounts to transforming to a coordinate system rotating with angular velocity Ω: the functions, f, k, l then transform to F, K, L.

Because of (4.9) and (4.10) only two of (4.8a, b, c) are independent. The following two combinations are of interest:

$$-2e^{\mu}\rho^{-1}k[(k + \Omega l)R_{00} + (f + \Omega^{2}l)R_{03} + \Omega(f - \Omega k)R_{33}]$$
$$= \left[\frac{1}{\rho}(FK_{\rho} - KF_{\rho})\right]_{\rho} + \left[\frac{1}{\rho}(FK_{z} - KF_{z})\right]_{z} = 0, \qquad (4.13a)$$

$$L_{\rho\rho} + L_{zz} - \frac{1}{\rho}L_{\rho} + \rho^{-2}L\Sigma = 8\pi mne^{\mu}F^{-1}(-FL - 2K^{2}), \quad (4.13b)$$

where $\Sigma \equiv F_{\mu}L_{\rho} + K_{\rho}^{2} + F_{z}L_{z} + K_{z}^{2}$. Equation (4.13b) is in fact (4.8c) multiplied by ρ. Equations (4.11a,b) for constant Ω imply

$$f - 2\Omega k - \Omega^{2}l = F = \text{constant} = F_{0}. \qquad (4.14)$$

Equation (4.13*a*) then implies

$$\Delta K \equiv K_{\rho\rho} + K_{zz} - \frac{1}{\rho}K_\rho = 0, \tag{4.15}$$

which can be solved as follows (see (2.26) and (2.27)):

$$K = \alpha\xi, \quad \Delta\xi = 0, \quad \xi = \rho\eta_\rho, \tag{4.16}$$

where η is a harmonic function:

$$\nabla^2\eta = \eta_{\rho\rho} + \eta_{zz} + \rho^{-1}\eta_\rho = 0, \tag{4.17}$$

and α is an arbitrary constant, which, if necessary, can be absorbed into the definition of ξ. Equation (4.13*b*) then reduces to the following equation:

$$\Delta L + \alpha^2\rho^{-2}L(\xi_\rho^2 + \xi_z^2) = 8\pi mn(-L - 2\alpha^2 F_0^{-1}\xi^2)e^\mu. \tag{4.18}$$

But

$$L = F^{-1}(\rho^2 - K^2) = F_0^{-1}(\rho^2 - \alpha^2\xi^2), \tag{4.19}$$

from which we get

$$\Delta L = -2F_0^{-1}\alpha^2(\xi_\rho^2 + \xi_z^2). \tag{4.20}$$

Substituting from (4.19), (4.20) into (4.18) we see that the latter reduces to

$$8\pi mn = \alpha^2\rho^{-2}(\xi_\rho^2 + \xi_z^2)e^{-\mu}, \tag{4.21}$$

which determines the number density n in terms of the function ξ and the function μ.

The function μ is given in terms of F, K, L and n by the following equations:

$$-e^\mu R_1^1 = -\tfrac{1}{2}(\mu_{\rho\rho} + \mu_{zz}) + \tfrac{1}{2}\rho^{-1}\mu_\rho + \tfrac{1}{2}\rho^{-2}(F_\rho L_\rho + K_\rho^2) = 4\pi mne^\mu, \tag{4.22a}$$

$$-e^\mu R_2^2 = -\tfrac{1}{2}(\mu_{\rho\rho} + \mu_{zz}) - \tfrac{1}{2}\rho^{-1}\mu_\rho + \tfrac{1}{2}\rho^{-2}(F_z L_z + K_z^2) = 4\pi mne^\mu, \tag{4.22b}$$

$$R_{12} = \tfrac{1}{2}\mu_z + \tfrac{1}{4}\rho^{-2}(F_\rho L_z + F_z L_\rho + 2K_\rho K_z) = 0. \tag{4.22c}$$

From (4.22*a, b*) we get

$$\mu_\rho = \tfrac{1}{2}\rho^{-1}(F_z L_z + K_z^2 - F_\rho L_\rho - K_\rho^2). \tag{4.23}$$

From (4.14), (4.16), (4.19), (4.22*c*) and (4.23) we get the following pair of equations for μ;

$$\mu_\rho = \tfrac{1}{2}\rho^{-1}\alpha^2(\xi_z^2 - \xi_\rho^2), \quad \mu_z = -\alpha^2\rho^{-1}\xi_\rho\xi_z, \tag{4.24}$$

the consistency of which is guaranteed by the fact that $\Delta\xi = 0$. Thus (4.14), (4.16), (4.17), (4.19), (4.21) and (4.24) give a class of exact solutions in terms of the function ξ satisfying $\Delta\xi = 0$ which can be expressed in terms of a harmonic function as in (4.16). This class of solutions is the Van Stockum interior solution for rigidly rotating dust.

It seems unlikely that the Van Stockum interior solution obtained above can represent the interior of a bounded distribution of dust which can match smoothly onto an asymptotically flat solution. This is what one would expect from a simple Newtonian analysis. Consider a possible bounded distribution of rotating dust and let us fix our attention on the upper extremity of the matter at the 'pole' P (Fig. 4.1). The rest of the matter is underneath it so that the resultant gravitational pull on the matter at P is downwards. However, since the pressure is zero, the gradient of the pressure is zero so there are no pressure forces and there is nothing to prevent the matter at P from collapsing onto the z-plane. The only possibility is that of an infinitesimally thin disc, for which there are no 'interior' solutions. This analysis cannot necessarily be taken over exactly to the general relativistic case, for in the latter situation, as remarked earlier, there are additional forces due to the rotation of matter which might conceivably prevent the matter at P from collapsing onto the z-plane. However, it seems unlikely that a bounded source of dust such as that in Fig. 4.1 exists in general relativity. The relativistic disc is a separate problem which has, for example, been considered by Bardeen and Wagoner (1969) and Lynden-Bell and Pineault (1978).

4.3. Rigidly rotating dust cylinder

The Van Stockum solution can be applied to the case of a rigidly rotating dust cylinder by specializing the function ξ in (4.16) to be independent of z. In this case the equation $\Delta \xi = 0$ can be solved trivially to yield

$$\xi = a\rho^2 + \xi_0. \tag{4.25}$$

Fig. 4.1. Illustration of the forces on the pole P of a bounded rotating distribution of matter.

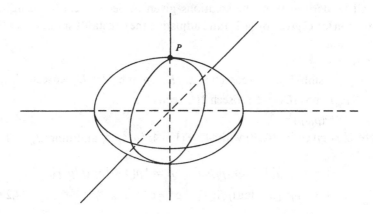

where a and ξ_0 are constants. We can set $a = 1$ without loss of generality since ξ is always multiplied by the arbitrary constant α.

Van Stockum was able to match smoothly the cylindrically symmetric solution obtained from (4.25) with the cylindrically symmetic form of the exterior solutions found in Section 2.6. We will not give the details of this matching procedure, but state the results in the form given by Bonnor (1980a), who obtained some additional results. The interior solution is given uniquely by a single parameter, whereas the exterior involves three different cases, depending on the mass per unit length of the cylinder. Bonnor calls these case I (low mass), case II (null case) and case III (ultrarelativistic). The exterior solutions here correspond to the cases of Section 2.6 given by $4AB + 1 > 0$ (case I), $4AB + 1 = 0$ (case II) and $4AB + 1 < 0$ (case III).

The interior solution (say for $\rho < \rho_0$, so that $\rho = \rho_0$ is the boundary of the cylinder) is given as follows:

$$F = 1, \quad K = \alpha\rho^2, \quad L = \rho^2(1 - \alpha^2\rho^2), \quad e^\mu = e^{-\alpha^2\rho^2}, \quad n = (2\pi)^{-1}\alpha^2 e^{\alpha^2\rho^2}$$
$$(4.26)$$

The function K has been obtained from (4.16) and (4.25) by setting $a = 1$ and $\xi_0 = 0$, which can be done without loss of generality. Both this choice and the choice $F_0 = 1$ can be made by a trivial change in the coordinate system. Similarly we get L from (4.19). The function μ is obtained from integrating the first relation in (4.24) and adjusting an arbitrary constant suitably.

Note that the exterior solution is given by an arbitrary harmonic function, U, which in the cylindrically symmetric case is given by

$$U = A'\log\rho + B' = A''\log(\rho/\rho_0), \qquad (4.27)$$

where A', B', A'', ρ_0 are constants, with $B' = -A''\log\rho_0$. The three cases mentioned earlier are given by the following three exterior solutions, which can all be derived from the solutions given in Section 2.6 by using the expression for U given by (4.27) and adjusting the constants suitably (see the next section.

Case I, $\alpha\rho_0 < \frac{1}{2}$

$$F = (\rho/\rho_0)\sinh(U_0 - U)\operatorname{cosech}U_0, \quad K = \rho\sinh(U + U_0)\operatorname{cosech}2U_0,$$
$$L = \tfrac{1}{2}\rho_0\rho\sinh(3U_0 + U)\operatorname{cosech}2U_0\operatorname{sech}U_0,$$
$$e^\mu = e^{-\alpha^2\rho_0^2}(\rho_0/\rho)^{2\alpha^2\rho_0^2}, \qquad (4.28a)$$

where U is given by (4.27) with $A'' = (1 - 4\alpha^2\rho_0^2)^{1/2}$, and $\tanh U_0 = A''$.

Case II, $\alpha\rho_0 = \frac{1}{2}$

$$F = (\rho/\rho_0)[1 - \log(\rho/\rho_0)], \quad K = \tfrac{1}{2}\rho[1 + \log(\rho/\rho_0)],$$
$$L = \tfrac{1}{4}\rho\rho_0[3 + \log(\rho/\rho_0)], \quad e^\mu = e^{-1/4}(\rho_0/\rho)^{1/2}. \qquad (4.28b)$$

Case III, $\alpha\rho_0 > \frac{1}{2}$

$$F = (\rho/\rho_0)\sin(U_0 - U)\operatorname{cosec} U_0, \quad K = \rho\sin(U + U_0)\operatorname{cosec} 2U_0,$$

$$L = \tfrac{1}{2}\rho_0\rho\sin(3U_0 + U)\operatorname{cosec} 2U_0\sec U_0, \quad e^\mu = e^{-\alpha^2\rho_0^2}(\rho_0/\rho)^{2\alpha^2\rho_0^2},$$

$$(4.28c)$$

where U is given by (4.27) with $A'' = (4\alpha^2\rho_0{}^2 - 1)^{1/2}$, and $\tan U_0 = A''$, $(0 < U_0 < \pi/2)$.

Here we have replaced f, k, l of Section 2.6 by F, K, L by transforming to a rotating coordinate system (see (4.12)).

The exterior solutions given by (4.28a, b, c) match smoothly onto the interior solution (4.26) at $\rho = \rho_0$, that is, the metric functions and their derivatives are all continuous at the boundary $\rho = \rho_0$. The mass and angular momentum per unit z coordinate are $m = \tfrac{1}{2}\alpha^2\rho_0^2$ and $j = \tfrac{1}{4}\alpha^3\rho_0^4$, respectively (Bonnor, 1980a).

It was stated by Frehland (1971) that the exterior metric is static in that it can be diagonalized (that is, transformed to a different system in which $K = 0$) (see also Som, Teixeira and Wolk, 1976). Bonnor (1980a) showed that this is not correct, because although case I can be diagonalized, the other two cases cannot. Bonnor showed this by proving that a hypersurface-orthogonal time-like Killing vector (see Section 1.8) exists in case I, but not in cases II or III. Even in case I, Bonnor shows that the diagonalization can be accomplished by the introduction of a periodic time coordinate (see Tipler, 1974). For this reason Bonnor calls this solution locally static, but not globally static, because in this case there does not exist a time coordinate which is a monotonically increasing function on future-pointing time-like curves.

An unusual feature of the Van Stockum solution is that in case III closed time-like curves occur in the exterior metric. These curves are given by

$$x^0 = t = \text{constant}, \quad x^1 = \rho = \text{constant}, \quad x^2 = z = \text{constant}, \quad x^3 = \phi,$$

$$(4.29)$$

because L given by (4.28c) is negative so the curves given by (4.29) are time-like and if ϕ has the usual interpretation as a periodic coordinate with range $0 < \phi < 2\pi$ and $\phi = \phi_0$ being identified with $\phi = \phi_0 + 2\pi$, then these curves are closed. The Kerr (1963) and Gödel (1949) solutions have closed time-like curves but, as noted by Bonnor, these solutions have uncertain sources whereas the Van Stockum solution arises from a reasonable form of matter, namely, dust. However, the infinite length of the cylinder possibly has some unphysical effects – for example, the metric is not asymptotically flat. Nevertheless, the possibility cannot be ruled out that in general relativity a very long cylinder of matter gives rise to some unusual behaviour.

4.4. Further remarks about the dust cylinder

In this section we shall indicate how the exterior solutions of the last section are derived from those of Section 2.6 and also derive some results for the interior solution (4.26) for the dust cylinder. We shall give here the derivation of (4.28a) from (2.58a, b, c). One can in a similar manner derive (4.28b, c) from (2.52a, b, c) and (2.48a, b, c). In (2.58a, b, c) we replace U by $U + 3U_0$ where U_0 is the constant in (4.28a), which is permissible since $U + 3U_0$ is also harmonic. We then choose U to be the cylindrically symmetric harmonic function given by (4.27). Next we transform (2.58a, b, c) to a rotating coordinate system given by (see (4.12))

$$L = l = (u_0 A)^{-1} \rho \sinh (U + 3U_0), \tag{4.31a}$$

$$K = k + \omega l = (u_0 A)^{-1} \rho [(\omega - (2A)^{-1}) \sinh (U + 3U_0) \\ + u_0 A \cosh (U + 3U_0)], \tag{4.31b}$$

$$F = f - 2\omega k - \omega^2 l = (u_0 A)^{-1} \rho [(-u_0^2 A^2 \\ - (\omega - (2A)^{-1})^2) \sinh (U + 3U_0) \\ + 2u_0 A ((2A)^{-1} - \omega) \cosh (U + 3U_0)]. \tag{4.31c}$$

If we now choose u_0, A, ω as follows:

$$u_0 A = 2\rho_0^{-1} \sinh 2U_0 \cosh U_0, \tag{4.32a}$$

$$\omega - (2A)^{-1} = -u_0 A \coth 2U_0. \tag{4.32b}$$

then the L, K and F given by (4.31a, b, c) are precisely of the form (4.28a).

We now derive some properties of the interior of the dust cylinder following Van Stockum (1937). What follows is a somewhat extended and modified form of Van Stockum's work. We started off with the metric (4.5) in which the components of the four-velocity u^μ are given by (4.6). We then transformed to the rotating system given by (4.12), which can be achieved by the introduction of a new angular coordinate ϕ' given by $\phi' = \phi - \Omega t$, where Ω is the constant angular velocity. By the transformation law of a contravariant vector we find that the transformed components u'^μ of u^μ are

$$u'^0 = u^0, \quad u'^1 = u^1 = 0, \quad u'^2 = u^2 = 0, \\ u'^3 = u^3 - \Omega u^0 = 0. \tag{4.33}$$

Thus the dust is at rest with respect to the rotating coordinate system since $u'^3 = 0$. We are interested in finding the angular velocity of the cylinder. We do this by defining the angular velocity of the cylinder to be the angular velocity relative to a non-rotating observer on the axis of symmetry. A non-rotating observer is best defined by the fact the unit tangent vectors in the direction of the space axes fixed with respect to the observer are Fermi–Walker transported (Walker, 1935) along the world-line of the observer (see

Misner *et al.*, 1973, Section 40.7). For the present it is sufficient to note that if the world-line is a geodesic, as is the case here (see below), the Fermi–Walker transport reduces to parallel transport (see (1.42)).

We transform to a coordinate system S' rotating with respect to the system S given by (F, K, L) (see (4.12) and (4.26)) with angular velocity Ω' so that the new values (F', K', L') are given by

$$L' = L, \quad K' = K + \Omega'L, \quad F' = F - 2\Omega'K - \Omega'^2 L. \qquad (4.34)$$

The four-velocity vector of an observer at rest in the system S' is given as follows:

$$\frac{dx^\mu}{ds} = (F'^{-1/2}, 0, 0, 0). \qquad (4.35)$$

It is readily verified that this worldline is a geodesic if it is situated on the axis of symmetry at $\rho = 0$, since then it satisfies the geodesic equation

$$\frac{d^2 x^\mu}{ds^2} + \Gamma^\mu_{\nu\sigma} \frac{dx^\nu}{ds} \frac{dx^\sigma}{ds} = 0. \qquad (4.36)$$

Equation (4.36) reduces to

$$\frac{d}{ds}(F'^{-1/2}, 0, 0, 0) + (\Gamma^0_{00}, \Gamma^1_{00}, \Gamma^2_{00}, \Gamma^3_{00}) F'^{-1} = 0 \qquad (4.37)$$

which can be seen to be satisfied at $\rho = 0$, if we evaluate the Γ^μ_{00} in S' and use the fact that $F'_\rho = 0$ for $\rho = 0$.

Next consider the unit tangent vector to the ρ-lines in S', which are defined by

$$(t = \text{constant}, \rho, z = \text{constant}, \phi = \text{constant}). \qquad (4.38)$$

This unit tangent is given by ξ^ν, where

$$\xi^\nu = (0, e^{-(1/2)\mu}, 0, 0). \qquad (4.39)$$

Then ξ^ν must be transported parallelly along dx^μ/ds, that is,

$$\frac{\partial \xi^\nu}{\partial x^\lambda} \frac{dx^\lambda}{ds} + \Gamma^\nu_{\lambda\sigma} \xi^\lambda \frac{dx^\sigma}{ds} = 0. \qquad (4.40)$$

With the use of (4.35), (4.39) and (1.98), equation (4.40) becomes

$$\rho^{-2}(F'K'_\rho - K'F'_\rho)F'^{-1/2} e^{-(1/2)\mu} = 0, \qquad (4.41a)$$

$$\rho^{-2}(K'K'_\rho + L'F'_\rho)F'^{-1/2} e^{-(1/2)\mu} = 0, \qquad (4.41b)$$

With the use of (4.34), equations (4.41a, b) become

$$\rho^{-2}[(FK_\rho - KF_\rho) + \Omega'(FL_\rho - LF_\rho) + \Omega'^2(LK_\rho - KL_\rho)]F'^{-1/2} e^{-(1/2)\mu} = 0, \qquad (4.42a)$$

$$\rho^{-2}[KK_\rho + LF_\rho + \Omega'(KL_\rho - LK_\rho)]F'^{-1/2}e^{-(1/2)\mu} = 0 \quad (4.42b)$$

We now substitute the expressions (4.26) for F, K, L in (4.42a, b) to get the following two equations:

$$\rho(1 - \alpha\Omega'\rho^2) = 0, \quad (4.43a)$$

$$\rho^{-1}(\Omega' + \alpha) - 2\alpha^2\rho\Omega' + \alpha^3\rho^3\Omega'^2 = 0. \quad (4.43b)$$

These are satisfied at $\rho = 0$ only if

$$\Omega' = -\alpha. \quad (4.44)$$

One can show in a similar manner that the unit tangent vector to the z-lines is propagated parallelly along the world-line of the observer on the axis of symmetry. We therefore obtain the metric in the non-rotating system of coordinates S', which is given as follows, with the use of (4.26), (4.34) and (4.44):

$$\left.\begin{array}{l} F' = 1 + \alpha^2\rho^2 + \alpha^4\rho^4, \quad K' = \alpha^3\rho^4, \\ L' = \rho^2(1 - \alpha^2\rho^2), \quad e^\mu = e^{-\alpha^2\rho^2}. \end{array}\right\} \quad (4.45)$$

From (4.44) it is evident that the angular velocity ω of the cylinder is given by

$$\omega = \alpha. \quad (4.46)$$

(Recall that we are using units in which the velocity of light and Newton's gravitational constant are unity.) From (4.26) we see that the density on the axis of symmetry, mn_0, is given by

$$2\pi mn_0 = \alpha^2, \quad (4.47)$$

so that the angular velocity is related to this density by

$$\omega = \alpha = (2\pi mn_0)^{1/2}. \quad (4.48)$$

Consider now a Newtonian cylinder of dust rotating rigidly with angular velocity ω'. If V is the gravitational potential, then balancing the centrifugal force with the gravitational force we get (ρ is distance from the axis of symmetry)

$$\rho\omega'^2 = V_\rho \quad (4.49)$$

which, together with the Poisson equation (m, n being as before the mass of particles and number density)

$$\nabla^2 V = V_{\rho\rho} + \rho^{-1}V_\rho = 4\pi mn, \quad (4.50a)$$

implies that n is constant and given by

$$\omega' = (2\pi mn)^{1/2}. \quad (4.50b)$$

Note the similarity between (4.48) and (4.50b), the difference being that in the Newtonian case (4.50*b*), *n* is necessarily constant whereas in the relativistic case (4.48) ω is determined by the value of *n* on the axis of symmetry $\rho = 0$. If the density on the axis of symmetry is that of water, then the period of rotation of the relativistic cylinder is approximately 2 hours and 42 minutes. Recalling that $\rho = \rho_0$ is the boundary of the cylinder, we see from (4.26) or (4.45) that we must have

$$\alpha\rho_0 < 1, \quad \text{or} \quad \omega\rho_0 < 1 \tag{4.51}$$

for otherwise the coefficient of $d\phi^2$ in the metric is positive in the interior of the cylinder. Since the velocity of light is unity, we see from (4.51) that the upper limit of ρ_0 is the same as the upper limit of the radius of a rotating cylinder in the special theory of relativity. However, ρ_0 is not the radius of the cylinder but is related to it as follows:

$$\rho_c = \int_0^{\rho_0} e^{-(1/2)\alpha^2\rho^2} d\rho. \tag{4.52}$$

If the density on the axis is again that of water the maximum radius is approximately 3.5×10^8 km.

We now consider the angular velocity with which a particle must be endowed to describe a ϕ-line (that is, a line with $\rho = $ constant, $z = $ constant) so that its world-line is a geodesic. This can be worked out by using the form (1.105*a*, *b*, *c*) of the geodesic equations, which are applicable here because of cylindrical symmetry. We set $d\rho/ds = 0$ in (1.105*a*, *b*, *c*) to get the following relations (we use the frame S so that the metric functions are given by (4.26)):

$$F\left(\frac{dt}{ds}\right)^2 - 2K\frac{dt}{ds}\frac{d\phi}{ds} - L\left(\frac{d\phi}{ds}\right)^2 = 1, \tag{4.53a}$$

$$\frac{d^2t}{ds^2} = 0, \quad \frac{d^2\phi}{ds^2} = 0. \tag{4.53b}$$

By adjusting constants suitably in the integration of (4.53*b*) we can take $t = s$, $d\phi/ds = d\phi/dt = \phi_0 = $ constant, without loss of generality. Substituting for F, K and L in (4.53*a*) from (4.26) we then obtain the following two values of ϕ_0, the angular velocity of a particle describing a ϕ-line:

$$\phi_0 = 0, \quad \phi_0 = -2\alpha(1 - \alpha^2\rho^2)^{-1}. \tag{4.54}$$

The first root confirms the fact that the dust particles follow geodesics (recall that (4.26) describes the rest frame of the cylinder). The second root gives the angular velocity (for any constant value of ρ) with which a particle must be endowed in order to traverse a thin, hollowed-out tube along a ϕ-line in a sense contrary to the sense of rotation of the cylinder. This angular velocity is, of course, with respect to the rest frame of the cylinder.

We refer to the interesting papers by Van Stockum (1937) and Bonnor (1980a) for more results on this problem.

4.5. Differentially rotating dust

In the case of axisymmetric differential rotation Ω is a function of ρ and z. Equations (4.8a,b,c), (4.9), (4.10) and (4.11a,b) are all valid in this case, with $\Omega = \Omega(\rho, z)$. We again transform to the new function L, K, F given (4.12) but now these functions no longer refer to a rotating coordinate system since Ω is not constant. From (4.11a, b) we get

$$F_\rho + 2\Omega_\rho K = 0, \tag{4.55a}$$
$$F_z + 2\Omega_z K = 0, \tag{4.55b}$$

from which it follows that

$$\Omega_\rho F_z - \Omega_z F_\rho = 0, \tag{4.56}$$

which implies that F is a function of Ω,

$$F = F(\Omega). \tag{4.57}$$

Equations (4.55a, b) and (4.57) then imply that K is also a function of Ω given by

$$F' = -2K \tag{4.58}$$

where the prime denotes differentiation with respect to Ω. (Note the use of the prime in this section as distinct from that of the last section.) Note that only two of (4.8a, b, c) are independent. From these equations one can derive the following equation:

$$K\Delta F - F\Delta K + 4K(K_\rho\Omega_\rho + K_z\Omega_z) - 2KL(\Omega_\rho^2 + \Omega_z^2)$$
$$+ (2K^2 + LF)\Delta\Omega + 2F(L_\rho\Omega_\rho + L_z\Omega_z) = 0. \tag{4.59}$$

With the use of $FL + K^2 = \rho^2$ (see 4.10)) and the identity

$$\Delta F = F'\Delta\Omega + F''(\Omega_\rho^2 + \Omega_z^2) \tag{4.60}$$

and similar identities for K, (4.34) can be written as follows:

$$(KF' - FK' + \rho^2 + K^2)\Delta\Omega$$
$$+ [KF'' - FK'' - F^{-1}F'(\rho^2 - K^2)](\Omega_\rho^2 + \Omega_z^2) + 4\rho\Omega_\rho = 0. \tag{4.61}$$

Let us now assume that the function Ω is given implicitly by ξ satisfying $\Delta\xi = 0$ and ρ^2 as follows:

$$\xi = \xi(\rho, \Omega), \tag{4.62}$$

and let

$$\dot{\xi} = \left(\frac{\partial \xi}{\partial \rho}\right)_{\Omega = \text{constant}}, \quad \xi' = \left(\frac{\partial \xi}{\partial \Omega}\right)_{\rho = \text{constant}} \tag{4.63}$$

Then

$$\xi_\rho = \dot{\xi} + \xi'\Omega_\rho, \tag{4.64a}$$

$$\xi_{\rho\rho} = \ddot{\xi} + 2\dot{\xi}'\Omega_\rho + \xi''\Omega_\rho^2 + \xi'\Omega_{\rho\rho}, \tag{4.64b}$$

$$\xi_{zz} = \xi''\Omega_z^2 + \xi'\Omega_{zz}. \tag{4.64c}$$

The equation $\Delta\xi = 0$ can then be written as

$$\Delta\xi = \xi'\Delta\Omega + \xi''(\Omega_\rho^2 + \Omega_z^2) + 2\dot{\xi}'\Omega_\rho + \ddot{\xi} - \rho^{-1}\dot{\xi} = 0. \tag{4.65}$$

Compare (4.65) with (4.61) multiplied by F^{-1}. For these two equations to coincide we must have

$$\xi' = F^{-1}(KF' - FK' + \rho^2 + K^2), \tag{4.66a}$$

$$\xi'' = F^{-1}(KF'' - FK'' - F^{-1}F'(\rho^2 - K^2)), \tag{4.66b}$$

$$2\dot{\xi}' = 4\rho F^{-1}, \tag{4.66c}$$

$$\ddot{\xi} - \rho^{-1}\dot{\xi} = 0. \tag{4.66d}$$

It is readily verified that (4.66a, b, c, d) are satisfied by the followng ξ:

$$\xi = \int^\Omega (F^{-1}KF' - K' + \rho^2 + K^2)\,d\Omega, \tag{4.67}$$

which gives Ω implicitly as a function of ρ^2 and ξ. Note that this solution involves one arbitrary function, which can be taken as $F(\Omega)$ (or $\Omega(F)$). The function K is then determined by (4.58), and Ω is then given as a function of ρ^2 and ξ by (4.67). This solution was first obtained by Winicour (1975). The corresponding function μ can be obtained from the following two relations:

$$R_{11} - R_{22} = \rho^{-1}\mu_\rho + \tfrac{1}{2}\rho^{-2}[F_\rho L_\rho + K_\rho^2 - F_z L_z - K_z^2$$
$$+ 2(KL_\rho - LK_\rho)\Omega_\rho - 2(KL_z - LK_z)\Omega_z + L^2(\Omega_\rho^2 - \Omega_z^2)] = 0, \tag{4.68a}$$

$$R_{12} = \tfrac{1}{2}\rho^{-1}\mu_z + \tfrac{1}{4}\rho^{-2}[F_\rho L_z + F_z L_\rho + 2K_\rho K_z + 2(KL_z - LK_z)\Omega_\rho$$
$$+ 2(KL_\rho - LK_\rho)\Omega_z + 2L^2\Omega_\rho\Omega_z] = 0, \tag{4.68b}$$

and the number density can be obtained from the following equation, which is derived from (4.8a, b, c):

$$8\pi mne^\mu = F^{-1}\Delta F + 2F^{-1}(2K' - L)(\Omega_\rho^2 + \Omega_z^2)$$
$$+ 2F^{-1}K\Delta\Omega + \rho^{-2}\Sigma, \tag{4.69}$$

where

$$\Sigma \equiv F_\rho L_\rho + F_z L_z + K_\rho^2 + K_z^2 + 2\Omega_\rho(KL_\rho - LK_\rho)$$
$$+ 2\Omega_z(KL_z - LK_z) + L^2(\Omega_\rho^2 + \Omega_z^2). \tag{4.69a}$$

Vishveshwara and Winicour (1977) have considered the cylindrically symmetric form of the above solution and the global solution for a differentially rotating dust cylinder and applied matching conditions at the boundary to exterior solutions derived in Section 2.6 (the cylindrically symmetric form of these). They discuss asymptotic behaviour at infinity and derive some interesting results (see also Hoenselaers and Vishveshwara, 1979*a*, *b*, and Chakraborty, 1980).

5

Rotating Einstein–Maxwell fields

5.1. Introduction

In this chapter we shall be concerned with the exterior gravitational and electromagnetic fields of rotating charged sources. The electromagnetic field has energy stored in it and hence contributes to the energy-momentum tensor in the region exterior to the sources. We will not be concerned in this chapter with equations satisfied by the sources but consider only some general properties of the sources as reflected in the exterior field. All the solutions we mention in this chapter are exterior solutions of the Einstein–Maxwell (EM) equations. These solutions are also known as 'electrovac' solutions. A great deal of work has been done on the EM equations. In this chapter we shall mainly be concerned with some general classes of solutions and the physical property of the rotating sources that these exterior solutions reflect.

Papapetrou (1947) and Majumdar (1947) independently discovered electrostatic (non-rotating) solutions of the EM equations which are given in terms of a single harmonic function. These solutions have no spatial symmetry (i.e. they are non-axisymmetric), and are produced by sources with $m = |e|$, m and e being the mass and charge respectively in suitable units. We call these the PM solutions (these solutions are distinct from the Papapetrou solutions discussed in Section 2.5). Weyl's (1917) electrostatic (non-rotating) solutions of the EM equations (these are distinct from the Weyl solutions of Section 2.3 – in this chapter we shall always refer to the electrostatic solutions) have axial symmetry, but the sources satisfy $m = \beta e$ where β is a constant, the same for all masses. These Weyl solutions are given in terms of a single axisymmetric harmonic function. Perjés (1971) and independently Israel and Wilson (1972) generalized the PM solutions to a class of stationary (rotating) solutions of the EM equations with no spatial symmetry which arise from sources satisfying

$$m = |e|, \quad \mathbf{h} = \pm \mu, \tag{5.1}$$

where **h** and μ are the angular momentum and magnetic moment vectors respectively of the source. We call these the PIW solutions. The PIW solutions are expressible in terms of two harmonic functions. One of the problems we will discuss in this chapter is the following one. There should exist a class of axially symmetric stationary (rotating) solutions of the EM equations which is related to the PIW class in a similar manner to that in which the Weyl class is related to the PM class. These solutions, which we will refer to as the class S of solutions, would be associated with sources satisfying the following relations:

$$m = \beta e, \quad \mathbf{h} = \beta' \boldsymbol{\mu}, \tag{5.2}$$

where β' is another constant. It was in an unsuccessful attempt to find this class of solutions that Bonnor (1973) found a class of axisymmetric rotating solutions of the EM equations that is expressible in terms of a single harmonic function. The Bonnor class is unphysical in a similar sense to that in which the Papapetrou class of Section 2.5 is unphysical, namely an asymptotically flat solution can be got only by making the mass of the source zero.

The general axisymmetric stationary exterior EM equations can be reduced to four coupled equations for four unknowns. The equations governing the class S reduce to two coupled equations for two unknowns (Islam, 1978b). In the process of deriving this result we will consider Ernst's (1968b) form of the EM equations, which is a generalization of his formulation of the Einstein equations considered in Chapter 3.

5.2. The field equations

The Einstein–Maxwell exterior equations, in suitable units, are as follows:

$$R_{\mu\nu} = 8\pi E_{\mu\nu} = -2F_\mu{}^\alpha F_{\nu\alpha} + \tfrac{1}{2} g_{\mu\nu} F_{\alpha\beta} F^{\alpha\beta}, \tag{5.3a}$$

$$F_{\mu\nu;\sigma} + F_{\nu\sigma;\mu} + F_{\sigma\mu;\nu} = 0, \tag{5.3b}$$

$$F^{\mu\nu}{}_{;\nu} = -4\pi J^\mu = 0, \quad F_{\mu\nu} = A_{\mu,\nu} - A_{\nu,\mu}, \tag{5.3c}$$

where $E_{\mu\nu}$ is the electromagnetic energy-momentum tensor and J^μ the four-current which we put equal to zero since we are here considering only the exterior field. $F_{\mu\nu}$ is the electromagnetic field tensor, defined in terms of the four-vector potential A_μ by (5.3c). A semicolon denotes covariant differentiation and a comma partial differentiation. Because of its definition in terms of A_μ, the tensor $F_{\mu\nu}$ satisfies (5.3b) identically. Equation (5.3a) follows from (1.35) if we interpret $T_{\mu\nu}$ as $E_{\mu\nu}$ and note that in this case the Ricci scalar R vanishes identically since $E_\mu{}^\mu \equiv 0$. It can also be shown, using standard procedure of Maxwell theory, that $E_{\mu\nu}$ has zero divergence:

$$E^{\mu\nu}{}_{;\nu} = 0, \qquad (5.4)$$

which represents the conservation of energy and momentum of the electromagnetic field.

We now specialize to the axially symmetric stationary metric considered earlier as follows:

$$ds^2 = f(dt - w d\phi)^2 - \rho^2 f^{-1} d\phi^2 - e^{\mu}(d\rho^2 + dz^2) \qquad (2.16)$$

(the use of this form will be justified in Chapter 6)
where f, w and μ are all functions of ρ and z. Writing $(x^0, x^1, x^2, x^3) = (t, \rho, z, \phi)$, the vector potential (A_0, A_1, A_2, A_3) can be written in terms of two scalar fields Φ and Φ' and the metric functions f and w as follows (Ernst, 1968b):

$$A_0 = \Phi, \quad A_1 = A_2 = 0, \quad \frac{\partial A_3}{\partial \rho} = w\Phi_\rho + \rho f^{-1}\Phi'_z, \quad \frac{\partial A_3}{\partial z} = w\Phi_z - \rho f^{-1}\Phi'_\rho,$$

$$(5.5)$$

where $\Phi_\rho \equiv \partial\Phi/\partial\rho$, etc. The consistency of the last two relations in (5.5) is guaranteed by (5.6d) below, as will be seen shortly. The field equations (5.3a) and (5.3c) in the metric (2.16) and for A_μ given by (5.5) yield, firstly, the following four equations:

$$f\nabla^2 f - f_\rho^2 - f_z^2 + \rho^{-2} f^4(w_\rho^2 + w_z^2) = 2f(\Phi_\rho^2 + \Phi_z^2 + \Phi_\rho'^2 + \Phi_z'^2), \qquad (5.6a)$$

$$f\Delta w + 2f_\rho w_\rho + 2f_z w_z = 4\rho f^{-1}(\Phi_z'\Phi_\rho - \Phi_\rho'\Phi_z), \qquad (5.6b)$$

$$f\nabla^2\Phi = f_\rho\Phi_\rho + f_z\Phi_z + \rho^{-1} f(w_z\Phi_\rho' - w_\rho\Phi_z'), \qquad (5.6c)$$

$$f\nabla^2\Phi' = f_\rho\Phi_\rho' + f_z\Phi_z' + \rho^{-1} f(w_\rho\Phi_z - w_z\Phi_\rho). \qquad (5.6d)$$

We remind the reader that $\nabla^2 \equiv \partial^2/\partial\rho^2 + \partial^2/\partial z^2 + \rho^{-1}\partial/\partial\rho$ and $\Delta \equiv \partial^2/\partial\rho^2 + \partial^2/\partial z^2 - \rho^{-1}\partial/\partial\rho$. Equations (5.6a, b) reduce to (2.12a, b) when $\Phi = \Phi' = 0$. It is readily verified that (5.6d) can be written as follows:

$$(w\Phi_\rho + \rho f^{-1}\Phi'_z)_z - (w\Phi_z - \rho f^{-1}\Phi'_\rho)_\rho = 0, \qquad (5.7)$$

which shows the consistency of the definition (5.5). Secondly, the field equations yield the following two equations for μ, the consistency of which is guaranteed by (5.6a–d):

$$\mu_\rho = -f^{-1}f_\rho + \tfrac{1}{2}\rho(f_\rho^2 - f_z^2) + 2\rho f^{-1}(\Phi_z^2 - \Phi_\rho^2 + \Phi_z'^2 - \Phi_\rho'^2)$$
$$+ \tfrac{1}{2}\rho^{-1}f^2(w_z^2 - w_\rho^2), \qquad (5.8a)$$

$$\mu_z = -f^{-1}f_z + \rho f_\rho f_z - 4\rho f^{-1}(\Phi_\rho\Phi_z + \Phi_\rho'\Phi_z') - \rho^{-1}f^2 w_\rho w_z. \qquad (5.8b)$$

Equations (5.8a, b) reduce to (2.13a, b) when $\Phi = \Phi' = 0$. There is another equation expressing $\mu_{\rho\rho} + \mu_{zz}$ in terms of the other functions, but this equation follows from the others. The proof is similar to that of the corresponding case in Chapter 2. The basic equations are thus (5.6a–d), since μ can be obtained by quadratures once f, w, Φ and Φ' have been solved from $(5.6a\text{--}d)$.

5.3. The Ernst form of field equations

Following Ernst ($1968b$) we shall transform $(5.6a\text{--}d)$ to two coupled complex equations. This will be useful for establishing a connection between solutions of the pure Einstein equations and certain kinds of solutions of the EM equations. Equation ($5.6b$) can be written as follows:

$$\frac{\partial}{\partial\rho}[\rho^{-1}f^2 w_\rho - 2(\Phi\Phi'_z - \Phi'\Phi_z)] + \frac{\partial}{\partial z}[\rho^{-1}f^2 w_z + 2(\Phi\Phi'_\rho - \Phi'\Phi_\rho)] = 0,$$

$$(5.9)$$

which implies the existence of a potential u' such that

$$\rho^{-1}f^2 w_\rho - 2(\Phi\Phi'_z - \Phi'\Phi_z) = u'_z, \quad \rho^{-1}f^2 w_z + 2(\Phi\Phi'_\rho - \Phi'\Phi_\rho) = -u'_\rho.$$

$$(5.10)$$

Eliminating w in favour of u' from $(5.6a\text{--}d)$, these equations get transformed to the following equations:

$$f\nabla^2 f - f_\rho^2 - f_z^2 = 2f(\Phi_\rho^2 + \Phi_z^2 + \Phi_\rho'^2 + \Phi_z'^2) - (2\Phi\Phi'_\rho - 2\Phi'\Phi_\rho + u'_\rho)^2$$
$$- (2\Phi\Phi'_z - 2\Phi'\Phi_z + u'_z)^2, \qquad (5.11a)$$

$$f\nabla^2 u = 4\Phi\Phi'(\Phi_\rho^2 + \Phi_z^2 - \Phi_\rho'^2 - \Phi_z'^2) + 4(\Phi'^2 - \Phi^2)(\Phi_\rho\Phi'_\rho + \Phi_z\Phi'_z)$$
$$- 2\Phi'(\Phi_\rho u'_\rho + \Phi_z u'_z) - 2\Phi(\Phi_\rho u'_\rho + \Phi_z u'_z) + 2(f_\rho u'_\rho + f_z u'_z)$$
$$- 2\Phi'(f_\rho\Phi_\rho + f_z\Phi_z) + 2\Phi(f_\rho\Phi'_\rho + f_z\Phi'_z), \qquad (5.11b)$$

$$f\nabla^2\Phi = f_\rho\Phi_\rho + f_z\Phi_z - 2\Phi(\Phi_\rho'^2 + \Phi_z'^2)$$
$$+ 2\Phi'(\Phi_\rho\Phi'_\rho + \Phi_z\Phi'_z) - u'_\rho\Phi'_\rho - u'_z\Phi'_z, \qquad (5.11c)$$

$$f\nabla^2\Phi' = f_\rho\Phi'_\rho + f_z\Phi'_z + 2\Phi(\Phi_\rho\Phi'_\rho + \Phi_z\Phi'_z)$$
$$- 2\Phi'(\Phi_\rho^2 + \Phi_z^2) + u'_\rho\Phi_\rho + u'_z\Phi_z. \qquad (5.11d)$$

Define the complex functions E and F as follows:

$$E = f - \Phi^2 - \Phi'^2 + iu', \quad F = \Phi + i\Phi'. \qquad (5.12)$$

Equations ($5.11a\text{--}d$) can then be combined into the following two complex equations:

$$(\operatorname{Re}E + |F|^2)\nabla^2 E = E_\rho^2 + E_z^2 + 2F^*(E_\rho F_\rho + E_z F_z), \qquad (5.13a)$$

$$(\operatorname{Re}E + |F|^2)\nabla^2 F = E_\rho F_\rho + E_z F_z + 2F^*(F_\rho^2 + F_z^2). \qquad (5.13b)$$

where $\operatorname{Re}E = f - \Phi^2 - \Phi'^2$ and a star denotes complex conjugation. We leave it as an exercise for the reader to show this. Equation ($5.13a$) reduces to (3.6) when $F = 0$. Equations ($5.13a,b$) give Ernst's formulation of the EM equations. The complex functions E and F are not uniquely determined by the metric and the electromagnetic field but can be subjected to the following gauge transformation:

$$E' = E - |C|^2 - 2C^*F + iC', \tag{5.14a}$$

$$F' = F + C, \tag{5.14b}$$

where C and C' are respectively a complex and real constant. That is E', F' given by (5.14*a, b*) satisfy the same equations as E and F, (5.13*a, b*). Clearly (5.13*a, b*) are also invariant under $F \to \exp(i\alpha) F$ where α is real.

We now consider a particular class of solutions of (5.13*a, b*) with the property that the function E is an analytic function of the complex variable F. This defines a subclass of the class of all solutions of the equations (5.13*a, b*). The reason for considering this subclass is that we can establish a correspondence between it and solutions of the pure Einstein equation (3.6). We substitute $E = E(F)$ in (5.13*a, b*) and see that the condition for the two equations to be equivalent is the following:

$$\frac{d^2 E}{dF^2} = 0, \tag{5.15}$$

so that E is a linear function of F. We now envisage a situation in which the electromagnetic field vanishes at infinity and the metric tends to that of Minkowski. That is, $E \to 1$ and $F \to 0$ at infinity. With these conditions the relation between E and F can be written as follows:

$$E = 1 - 2Q^{-1}F, \tag{5.16}$$

where $Q = q + iq'$ is a complex constant. Three cases arise according as to whether $|Q| < 1$, $= 1$ or > 1.

Case (i) $|Q| < 1$. Define ξ by

$$E = \frac{\alpha\xi - 1}{\alpha\xi + 1}, \quad \alpha = (1 - QQ^*)^{1/2}. \tag{5.17}$$

Then (5.13*a, b*) reduce to the single complex equation

$$(\xi\xi^* - 1)\nabla^2\xi = 2\xi^*(\xi_\rho^2 + \xi_z^2). \tag{5.18}$$

This is exactly the same as (3.8) of the pure Einstein case. Thus any solution of the stationary axisymmetric Einstein equations gives a corresponding solution of the EM equations through the above procedure.

Case (ii) $|Q| = 1$. In this case one can write $E = 1 - 2e^{-i\varepsilon}F$, where ε is a real constant. Carry out the gauge transformation (5.14*a, b*) with $C = -e^{i\varepsilon}$ and $C' = 0$. The result is

$$E' = E - 1 + 2e^{-i\varepsilon}F = 0, \tag{5.19a}$$

$$F' = F - e^{i\varepsilon} \tag{5.19b}$$

so that in the primed functions (5.13a) is satisfied identically and (5.13b) reduces to (since F'^* cancels)

$$F'\nabla^2 F' = 2(F_\rho'^2 + F_z'^2), \tag{5.20}$$

which is equivalent to

$$\nabla^2(F'^{-1}) = 0. \tag{5.21}$$

This leads to the axisymmetric form of the PIW solutions, as we shall see in the next section.

Case (iii) $|Q| > 1$. Write

$$\xi' = (1 - QQ^*)^{-1/2}(F - Q^{*-1}). \tag{5.22}$$

Then (5.13a, b) reduce to the following equation:

$$(\xi'\xi'^* + 1)\nabla^2 \xi' = 2\xi'^*(\xi_\rho'^2 + \xi_z'^2), \tag{5.23}$$

an equation which is similar to but distinct from (5.18). Thus in this case a correspondence with the pure Einstein equation cannot be found.

5.4. The Perjés–Israel–Wilson solutions

The axisymmetric form of these solutions is obtained from (5.21), which implies, together with the definition (5.12) of the function F (we drop the prime in (5.21)),

$$F^{-1} = (\Phi + i\Phi^{-1})^{-1} = \eta + i\zeta, \tag{5.24}$$

where η and ζ are independent real harmonic functions. From (5.24) and the fact that E is zero we get (see (5.12)) the following expressions for f, Φ and Φ':

$$f = \Phi^2 + \Phi'^2 = (\eta^2 + \zeta^2)^{-1}, \quad \Phi = (\eta^2 + \zeta^2)^{-1}\eta, \quad \Phi' = -(\eta^2 + \zeta^2)^{-1}\zeta. \tag{5.25}$$

Since $u' = 0$ (see (5.12)), the corresponding w can be gotten from (5.10) and (5.25) as follows:

$$w_\rho = 2\rho(\zeta\eta_z - \eta\zeta_z), \quad w_z = -2\rho(\zeta\eta_\rho - \eta\zeta_\rho), \tag{5.26}$$

the consistency of which is guaranteed by the fact that η and ζ are harmonic. Equation (5.25) and (5.26) give the axisymmetric form of the PIW solutions, which depend on two harmonic functions. As will be seen in a later section, this class contains a subclass that is asymptotically flat. By comparison with the pure Einstein case, this class is a relatively large class of asymptotically flat solutions. Thus asymptotic flatness seems easier to achieve for the EM equations than the pure Einstein equations. This reflects the fact that a greater variety of sources is available when one is considering charged sources.

We have given above the derivation of the axisymmetric form of the PIW solutions. We shall give here without derivation the non-axisymmetric form of the PIW solution. The metric is as follows:

$$ds^2 = |U|^{-2}(dt + \Omega^{(1)} dx + \Omega^{(2)} dy + \Omega^{(3)} dz)^2 - |U|^2(dx^2 + dy^2 + dz^2),$$
(5.27)

where U is any complex solution of Laplace's equation in the cartesian coordinates x, y, z (i.e. U is of the form $U_1 + iU_2$ where U_1 and U_2 are real non-axisymmetric harmonic functions), and the vector $\Omega = (\Omega^{(1)}, \Omega^{(2)}, \Omega^{(3)})$ is found by solving the equation

$$\nabla \wedge \Omega = i(U\nabla U^* - U^*\nabla U),$$
(5.28)

U^* being the complex conjugate of U. The electromagnetic field can be described in terms of non-axisymmetric potentials Φ, Φ' which are given by

$$\Phi + i\Phi' = U^{-1} = U^*/|U|^2.$$
(5.29)

To get the axisymmetric form of the solution we choose Ω, x, y as follows:

$$\Omega^{(3)} = 0, \quad \Omega^{(1)} = \rho^{-2}yw, \quad \Omega^{(2)} = -\rho^{-2}xw, \quad x = \rho \cos\phi, \quad y = \rho \sin\phi$$
(5.30)

and take $U = \eta + i\zeta$, where η and ζ are axisymmetric harmonic functions. Then the condition (5.28) yields the equations (5.26), with the use of $\partial\eta/\partial x = \rho^{-1}x\eta_\rho$, etc., and the metric (5.27) reduces to (2.16). The corresponding f, Φ, Φ' are also given by (5.25). From the form (5.27)–(5.29) of the PIW solutions one can get solutions with many charged black holes (Hartle and Hawking, 1972). Israel and Spanos (1973) have obtained an interior solution similar to the PIW solutions.

5.5. System of two equations for the class S of solutions

In this section we shall show that the reduction of the equations to two coupled equations by Ernst's method essentially gives the class S of solutions. We shall, however, carry out a slightly modified form of Ernst's method for this reduction and also indicate a separate method to give insight into the structure of the EM equations.

To obtain his solution Bonnor (1973) started with (5.6a–d) and the following assumptions about the form of the functions f, w and Φ':

$$f = f(\Phi), \quad \Phi' = \Phi'(\Phi), \quad w_\rho = -\rho s(\Phi)\Phi_z, \quad w_z = \rho s(\Phi)\Phi_\rho,$$
(5.31)

where f, Φ' and s are functions of Φ, to be determined. Consider the following more general assumptions about the form of f and w:

$$f = f(\Phi, \Phi'), \quad w_\rho = -\rho[s_1(\Phi, \Phi')\Phi_z + s_2(\Phi, \Phi')\Phi'_z],$$
$$w_z = \rho[s_1(\Phi, \Phi')\Phi_\rho + s_2(\Phi, \Phi')\Phi'_\rho], \tag{5.32}$$

that is, f is a function of Φ and Φ', and w is expressible in terms of Φ and Φ' in the manner shown in (5.32) where s_1 and s_2 are functions of Φ and Φ'. The form (5.32) reduces to (5.31) if Φ' is regarded as a function of Φ. It can be shown, after a lengthy calculation, that (5.32) leads to the following form for f, s_1 and s_2:

$$f = \Sigma + \varepsilon_0, \quad s_1 = 2(\Sigma + \varepsilon_0)^{-1}(\Phi' - \Phi'_0), \quad s_2 = -2(\Sigma + \varepsilon_0)^{-1}(\Phi - \Phi_0), \tag{5.33}$$

where $\Sigma \equiv (\Phi - \Phi_0)^2 + (\Phi' - \Phi'_0)^2$, and Φ_0, Φ'_0 and ε_0 are constants. The functions Φ, Φ' satisfy the following coupled equations:

$$\nabla^2\Phi = 2(\Sigma + \varepsilon_0)^{-1}[\Phi_\rho^2 + \Phi_z^2 - \Phi_\rho'^2 - \Phi_z'^2 + 2(\Phi' - \Phi'_0)(\Phi_\rho\Phi'_\rho + \Phi_z\Phi'_z)], \tag{5.34a}$$

$$\nabla^2\Phi' = 2(\Sigma + \varepsilon_0)^{-1}[\Phi_\rho'^2 + \Phi_z'^2 - \Phi_\rho^2 - \Phi_z^2 + 2(\Phi - \Phi_0)(\Phi_\rho\Phi'_\rho + \Phi_z\Phi'_z)]. \tag{5.34b}$$

We do not give the derivation of (5.33) and (5.34a, b) here as these equations can be derived much more simply from Ernst's formulation. In these equations Φ_0, Φ'_0 can be set equal to zero without loss of generality since the potentials Φ and Φ' are undefined to within additive constants, but we retain Φ_0, Φ'_0 here for the present.

Equation (5.13a) is satisfied identically by $E = $ constant, independent of what F is. Setting $\mathrm{Re}\, E = \varepsilon_1$, where ε_1 is a real constant, (5.13b) then reduces to the following equation:

$$(\varepsilon_1 + |F|^2)\nabla^2 F = 2F^*(F_\rho^2 + F_z^2). \tag{5.35}$$

Thus the set (E, F) where E is a constant and F is a solution of (5.35) constitutes a solution of the system (5.13a, b). The constant ε_1 can be taken to be real without loss of generality, for by the gauge transformation (5.14a, b) an imaginary part of E can be eliminated by a suitable choice of the constant C' in (5.14a). This amounts to saying that if the potential u' (see (5.10)) is constant (as is the case if E is constant), it is irrelevant what constant value u' has. Choosing E to be real amounts to setting $u' = 0$.

With F given by (5.12) it is readily verified that (5.34a, b) are equivalent to (5.35) if ε_0 is identified with ε_1 and $\Phi_0 = \Phi'_0 = 0$. Thus the reduction to two equations (5.34a, b) can be carried out by simply setting $E = $ constant in (5.13a, b). However, the derivation from (5.32) is interesting in that it shows that the equation (5.34a, b) or (5.35) follows under much more general assumptions and secondly it shows that the Bonnor solutions are a subclass of the solutions of (5.34*a*, *b*).

We assert that the set $(E = \varepsilon_1, F)$ (with ε_1 a real constant and F satisfying (5.35)) yields the class S of solutions. Before showing this we examine how the set $(E = \varepsilon_1, F)$ fits into Ernst's scheme for relating solutions of the Einstein and EM equations. A necessary condition for relating solutions of the EM and Einstein equations is that E and F should satisfy (5.16). Clearly $(E = \varepsilon_1, F)$ does not satisfy (5.16), for then F would also be a constant, yielding a trivial solution. But a gauge transformation of $(E = \varepsilon_1, F)$ does satisfy (5.16). To see this we carry out the transformation (5.14*a*, *b*) and make E', F' – the transformed functions – satisfy (5.16), that is, we set

$$E' = 1 - 2Q^{-1}F'. \tag{5.36}$$

This is equivalent to

$$\varepsilon_1 - |C|^2 - 2C^*F + iC' = 1 - 2Q^{-1}(F + C) \tag{5.37}$$

which yields the following relations:

$$\varepsilon_1 = 1 - |C|^2, \quad C' = 0, \quad C^* = Q^{-1}. \tag{5.38}$$

It is clear from (5.38) that if $\varepsilon_1 < 0$, one can find a gauge transformation such that E' and F' satisfy (5.36) with $|Q| < 1$. Thus in this case Ernst's scheme relates $(E = \varepsilon_1 < 0, F)$ with solutions of the pure Einstein equations in the form (5.18). The case $\varepsilon_1 = 0$ corresponds to $|Q| = 1$ and one gets the PIW solutions, as seen from case (ii) of Section 5.3 and from Section 5.4. It is also clear from (5.38) that $\varepsilon_1 > 0$ corresponds to $|Q| > 1$, so that $(E = \varepsilon_1 = > 0, F)$ correspond to solutions of (5.23) and therefore do not correspond to the pure Einstein equations. Thus the problem of finding the class S includes the problem of finding the general solution of Ernst's equation (5.18) (i.e., the general rotating solution of Einstein's equations).

We now show that $(E = \varepsilon_1, F)$ represents the class S. Recall that the class S of solutions are stationary, axisymmetric solutions of the EM equations that have sources with mass proportional to charge and angular momentum proportional to the magnetic moment. Thus the class S reduces to the Weyl electrostatic solutions in the absence of rotation (since in the Weyl solutions the mass is proportional to charge) and has the axisymmetric form of the PIW solutions as a subclass, since in the latter the sources have equal values of mass and charge, and of angular momentum and magnetic moment respectively. It is also clear from the work of Bonnor (1973), whose solution is given in Section 5.7, that the class reduces to his solution in a suitable limit.

We first show that in the absence of rotation $(E = \varepsilon_1, F)$ reduces to the electrostatic Weyl solution, just as in the absence of rotation the (non-axisymmetric) PIW solutions reduce to the electrostatic PM solutions. In the absence of rotation $\Phi' = 0$, and (5.35) gives

$$f = \varepsilon_1 + \Phi^2, \tag{5.39a}$$

$$(\varepsilon_1 + \Phi^2)\nabla^2\Phi = 2\Phi(\Phi_\rho^2 + \Phi_z^2). \tag{5.39b}$$

Equation (5.39b) has the solution that Φ is a function of the harmonic function ξ, the functional dependence of Φ on ξ being given by

$$\nabla^2\xi = 0, \quad \xi = \xi_0 \int^\Phi \frac{d\Phi}{\varepsilon_1 + \Phi^2}, \tag{5.40}$$

where ξ_0 is an arbitrary constant. Equations (5.39a) and (5.40) give Weyl's electrostatic solution.

Next we show that $(E = \varepsilon_1, F)$ has the axisymmetric PIW solution as a subclass. In fact we have already noted that $(E = 0, F)$ corresponds to (5.36) with $|Q| = 1$ and therefore yields the PIW solutions as seen by case (ii) and Section 5.4.

As is clear from (5.31) and (5.32) (this can be checked in detail) the Bonnor solutions are a subclass of $(E = \varepsilon_1, F)$. That the sources for $(E = \varepsilon_1, F)$ have masses proportional to the charge is suggested by the fact that it reduces, in the absence of rotation, to the electrostatic Weyl solutions which have this property. But this property of $(E = \varepsilon_1, F)$ and the property that the angular momentum is proportional to the magnetic moment, can be shown to hold by the expansion technique of the next section. Thus the solutions $(E = \varepsilon_1, F)$ have all the properties by which we have defined the class S.

5.6. Class of approximate solutions

In this section we shall derive a class of approximate solutions of the EM equations depending on four harmonic functions (Islam, 1976b). In the next section we shall consider the reduction of this solution to approximate forms of the Weyl electrostatic solution, the Bonner solution and the PIW solutions respectively. In Section 5.8 we shall consider physical interpretation of the approximate solution. There are several reasons for considering this problem in some detail. Firstly it gives some insight into the structure of the general solution of the EM equations. Secondly it gives some idea of the connection of the general solution to the variety of sources of the electromagnetic and gravitational field. In particular, it makes clear the connection of the various classes of solutions mentioned earlier in the chapter to the sources giving rise to these solutions.

For convenience we introduce the function h instead of f given by $h = f^{-1}$ just as we did for the Einstein equations in (2.61a,b). Equations (5.6a–d) then transform as follows:

$$h\nabla^2 h - h_\rho^2 - h_z^2 - \rho^{-2}(w_\rho^2 + w_z^2) = -2h^3(\Phi_\rho^2 + \Phi_z^2 + \Phi_\rho'^2 + \Phi_z'^2),$$
(5.41a)

$$h\Delta w - 2h_\rho w_\rho - 2h_z w_z = 4\rho h^3(\Phi_z'\Phi_\rho - \Phi_\rho'\Phi_z),$$
(5.41b)

$$h\nabla^2\Phi' = -h_\rho\Phi_\rho' - h_z\Phi_z' + \rho^{-1}(w_\rho\Phi_z - w_z\Phi_\rho),$$
(5.41c)

$$h\nabla^2\Phi = -h_\rho\Phi_\rho - h_z\Phi_z + \rho^{-1}(w_z\Phi_\rho' - w_\rho\Phi_z').$$
(5.41d)

Equations (5.8a, b) reduce to the following two:

$$h^2\mu_\rho = hh_\rho + \tfrac{1}{2}\rho(h_\rho^2 - h_z^2) + 2\rho h^3(\Phi_z^2 - \Phi_\rho^2 + \Phi_z'^2 - \Phi_\rho'^2)$$
$$+ \tfrac{1}{2}\rho^{-1}(w_z^2 - w_\rho^2),$$
(5.42a)

$$h^2\mu_z = hh_z + \rho h_\rho h_z - 4\rho h^3(\Phi_\rho\Phi_z + \Phi_\rho'\Phi_z') - \rho^{-1}w_\rho w_z.$$
(5.42b)

We assume that there is a solution of the field equations (5.41a–d) which depends smoothly on the parameter λ and a certain number of harmonic functions (see Section 2.7). We assume further that it is possible to expand the function h, w, Φ, Φ' in a power series in λ, as follows:

$$h = \sum_{n=0}^{\infty} \lambda^n h^{(n)}, \quad w = \sum_{n=0}^{\infty} \lambda^n w^{(n)}, \quad \Phi = \sum_{n=0}^{\infty} \lambda^n \Phi^{(n)}, \quad \Phi' = \sum_{n=0}^{\infty} \lambda^n \Phi'^{(n)}.$$
(5.43)

We also assume that the electromagnetic field vanishes and space–time becomes flat when $\lambda = 0$, so that $h^{(0)}, w^{(0)}, \Phi^{(0)}, \Phi'^{(0)}$ are constant, with $h^{(0)} = 1$, $w^{(0)} = 0$. Substituting (5.43) into (5.41a–d) and equating powers of λ we get, to first order, the equations:

$$\nabla^2 h^{(1)} = 0, \quad \Delta w^{(1)} = 0, \quad \nabla^2\Phi'^{(1)} = 0, \quad \nabla^2\Phi^{(1)} = 0,$$
(5.44)

so that

$$h^{(1)} = \tau, \quad \Phi^{(1)} = \xi, \quad \Phi'^{(1)} = \eta$$
(5.45)

where τ, ξ, and η are independent harmonic functions and the equation for $w^{(1)}$ can be solved as follows:

$$w_\rho^{(1)} = \rho\sigma_z, \quad w_z^{(1)} = -\rho\sigma_\rho,$$
(5.46)

where σ is a harmonic function. Equation (5.46) is equivalent to the solution $w^{(1)} = \rho\zeta_\rho$, where ζ is harmonic, which we used in (2.27), but (5.46) is preferable for the present. In fact $\sigma = -\zeta_z$.

In the second order we get the following four equations:

$$\nabla^2 h^{(2)} - h_\rho^{(1)2} - h_z^{(1)2} - \rho^{-2}(w_\rho^{(1)2} + w_z^{(1)2}) = -2(\Phi_\rho^{(1)2} + \Phi_z^{(1)2} + \Phi_\rho'^{(1)2}$$
$$+ \Phi_z'^{(1)2}),$$
(5.47a)

$$\Delta w^{(2)} - 2h_\rho^{(1)} w_\rho^{(1)} - 2h_z^{(1)} w_z^{(1)} = 4\rho(\Phi_z'^{(1)}\Phi_\rho^{(1)} - \Phi_\rho'^{(1)}\Phi_z^{(1)}), \tag{5.47b}$$

$$\nabla^2\Phi'^{(2)} = -h_\rho^{(1)}\Phi_\rho'^{(1)} - h_z^{(1)}\Phi_z'^{(1)} + \rho^{-1}(w_\rho^{(1)}\Phi_z^{(1)} - w_z^{(1)}\Phi_\rho^{(1)}), \tag{5.47c}$$

$$\nabla^2\Phi^{(2)} = -h_\rho^{(1)}\Phi_\rho^{(1)} - h_z^{(1)}\Phi_z^{(1)} - \rho^{-1}(w_\rho^{(1)}\Phi_z'^{(1)} - w_z^{(1)}\Phi_\rho'^{(1)}). \tag{5.47d}$$

Substitution of (5.45) and (5.46) into (5.47a–d) yields the following equations:

$$\nabla^2 h^{(2)} = \tau_\rho^2 + \tau_z^2 + \sigma_\rho^2 + \sigma_z^2 - 2(\xi_\rho^2 + \xi_z^2 + \eta_\rho^2 + \eta_z^2), \tag{5.48a}$$

$$\Delta w^{(2)} = 2\rho(\tau_\rho\sigma_z - \tau_z\sigma_\rho) + 4\rho(\xi_\rho\eta_z - \xi_z\eta_\rho), \tag{5.48b}$$

$$\nabla^2\Phi'^{(2)} = -\tau_\rho\eta_\rho - \tau_z\eta_z + \xi_\rho\sigma_\rho + \xi_z\sigma_z, \tag{5.48c}$$

$$\nabla^2\Phi^{(2)} = -\tau_\rho\xi_\rho - \tau_z\xi_z - \eta_\rho\sigma_\rho - \eta_z\sigma_z. \tag{5.48d}$$

Equations (5.48a, c, d) can be solved as follows:

$$h^{(2)} = \tfrac{1}{2}\tau^2 + \tfrac{1}{2}\sigma^2 - \xi^2 - \eta^2, \tag{5.49a}$$

$$\Phi'^{(2)} = -\tfrac{1}{2}\tau\eta + \tfrac{1}{2}\xi\sigma, \tag{5.49b}$$

$$\Phi^{(2)} = -\tfrac{1}{2}\tau\xi - \tfrac{1}{2}\eta\sigma. \tag{5.49c}$$

The solution of (5.48b) is given by

$$w_\rho^{(2)} = \rho(\tau\sigma_z - \sigma\tau_z) + 2\rho(\xi\eta_z - \eta\xi_z),$$
$$w_z^{(2)} = -\rho(\tau\sigma_\rho - \sigma\tau_\rho) - 2\rho(\xi\eta_\rho - \eta\xi_\rho), \tag{5.50}$$

the consistency of which is guaranteed by the fact that τ, σ, ξ, η are harmonic functions. We have ignored arbitrary harmonic functions that could be added to (5.49a, b, c) and an arbitrary solution of $\Delta w^{(2)} = 0$ that could be added to $w^{(2)}$; we will come back to this point. The equations (5.48a, c, d) have been solved by repeated application of the identity (2.72) and (5.48b) has been solved in a similar manner to which (2.74) was solved.

The equations for $h^{(3)}$, $w^{(3)}$, $\Phi^{(3)}$, $\Phi'^{(3)}$ after substitution for $h^{(1)}$, $w^{(1)}$, $\Phi^{(1)}$, $\Phi'^{(1)}$ from (5.45) and (5.46) and for $h^{(2)}$, $w^{(2)}$, $\Phi^{(2)}$, $\Phi'^{(2)}$ from (5.49a, b, c) and (5.50), and after some simplification, reduce to the following:

$$\begin{aligned}
\nabla^2 h^{(3)} = {}& \tau(\tau_\rho^2 + \tau_z^2) + \tau(\sigma_\rho^2 + \sigma_z^2) - 2\tau(\xi_\rho^2 + \xi_z^2 + \eta_\rho^2 + \eta_z^2) \\
& - 2\xi(\tau_\rho\xi_\rho + \tau_z\xi_z) - 2\eta(\tau_\rho\eta_\rho + \tau_z\eta_z) + 2\xi(\sigma_\rho\eta_\rho + \sigma_z\eta_z) \\
& - 2\eta(\sigma_\rho\xi_\rho + \sigma_z\xi_z),
\end{aligned} \tag{5.51a}$$

$$\begin{aligned}
\Delta w^{(3)} = {}& 2\rho[2\tau(\xi_\rho\eta_z - \xi_z\eta_\rho) + \tau(\tau_\rho\sigma_z - \tau_z\sigma_\rho) + \tau_\rho(\xi\eta_z - \eta\xi_z) \\
& - \tau_z(\xi\eta_\rho - \eta\xi_\rho) - \sigma_z(\xi\xi_\rho + \eta\eta_\rho) + \sigma_\rho(\xi\xi_z + \eta\eta_z)],
\end{aligned} \tag{5.51b}$$

$$\begin{aligned}
\nabla^2\Phi'^{(3)} = {}& \tfrac{1}{2}\eta(\tau_\rho^2 + \tau_z^2) + \tfrac{1}{2}\tau(\tau_\rho\eta_\rho + \tau_z\eta_z) - \xi(\tau_\rho\sigma_\rho + \tau_z\sigma_z) \\
& - \tfrac{3}{2}\sigma(\tau_\rho\xi_\rho + \tau_z\xi_z) - \tfrac{3}{2}\sigma(\sigma_\rho\eta_\rho + \sigma_z\eta_z) + 2\eta(\eta_\rho^2 + \eta_z^2) \\
& + 4\xi(\xi_\rho\eta_\rho + \xi_z\eta_z) - 2\eta(\xi_\rho^2 + \xi_z^2) - \tfrac{1}{2}\tau(\sigma_\rho\xi_\rho + \sigma_z\xi_z) \\
& - \tfrac{1}{2}\eta(\sigma_\rho^2 + \sigma_z^2),
\end{aligned} \tag{5.51c}$$

$$\nabla^2 \Phi^{(3)} = \tfrac{1}{2}\xi(\tau_\rho^2 + \tau_z^2) + \tfrac{1}{2}\tau(\tau_\rho\xi_\rho + \tau_z\xi_z) + \eta(\tau_\rho\sigma_\rho + \tau_z\sigma_z)$$
$$+ \tfrac{3}{2}\sigma(\tau_\rho\eta_\rho + \tau_z\eta_z) - \tfrac{3}{2}\sigma(\sigma_\rho\xi_\rho + \sigma_z\xi_z) + 2\xi(\xi_\rho^2 + \xi_z^2)$$
$$+ 4\eta(\xi_\rho\eta_\rho + \xi_z\eta_z) - 2\xi(\eta_\rho^2 + \eta_z^2) + \tfrac{1}{2}\tau(\eta_\rho\sigma_\rho + \eta_z\sigma_z)$$
$$- \tfrac{1}{2}\xi(\sigma_\rho^2 + \sigma_z^2). \tag{5.51d}$$

It does not seem possible to find closed form solutions for any of the equations (5.51a–d) in terms of the harmonic functions τ, σ, ξ, η. It is possible that such solutions do not exist. As in Section 2.7, at each stage of this approximation scheme one gets a Poisson equation for $h^{(n)}$, etc., in which the right hand side is given in terms of the lower order functions, which are known in principle. The equation for $w^{(n)}$ can be converted into Poisson's equation as follows. Suppose the equation at the nth stage is

$$\Delta w^{(n)} = F(\rho, z) \tag{5.52}$$

where F is known in terms of the lower order functions. Define $w'^{(n)}$ by $w^{(n)} = \rho w'^{(n)}_\rho$, then with the use of an equation similar to (2.81) it can be readily verified that $w'^{(n)}$ satisfies the following Poisson equation:

$$\nabla^2 w'^{(n)} = \int^\rho \rho'^{-1} F(\rho', z) \, d\rho'. \tag{5.53}$$

These equations can be solved at each state in terms of integral representations (1.4), so that assuming the convergence of the series, this shows in principle that a general solution must depend on at least four harmonic functions. However, this procedure may not necessarily lead to the most general solution. This is because at each stage of the approximation we have been looking for coefficients of the powers of λ which are polynomials in the harmonic functions and their derivatives. There may exist solutions with more general functional form. For example, the Herlt (1978) class of electrostatic solutions of the EM equations (these are distinct from the Weyl electrostatic solutions) which depends on a harmonic function does not appear to be included in these schemes. It is an interesting problem to find how the above scheme can be modified to include the Herlt class of solutions, that is, to generalize the above procedure so that to order λ^3 the approximate solution also reduces to the Herlt solution in a suitable limit (see next section).

Equations (5.51a–d) can be considered as a 'laboratory' in which to test the possibility of exact solutions in terms of harmonic functions. In this the following identify is useful:

$$\nabla^2(FGH) = FG\nabla^2 H + FH\nabla^2 G + GH\nabla^2 F + 2F(H_\rho G_\rho + H_z G_z)$$
$$+ 2H(F_\rho G_\rho + F_z G_z) + 2G(H_\rho F_\rho + H_z F_z), \tag{5.54}$$

for any three functions F, G and H. For example, let $\tau = \alpha\xi$ and $\sigma = \beta\eta$, where α, β are constants. In this case (5.51a) can be integrated if

$$\alpha(1 - \beta^2) - 3\beta = 0 \qquad (5.55a)$$

the solution being

$$h^{(3)} = \tfrac{1}{6}\alpha(\alpha^2 - 4)\xi^3 - \tfrac{1}{2}(\alpha + \beta)\xi\eta^2. \qquad (5.55b)$$

Equation (5.51b) is equivalent to $\Delta w^{(3)} = 0$, and hence soluble, if

$$3\alpha + \beta(\alpha^2 - 1) = 0. \qquad (5.56)$$

Equation (5.51c) is soluble explicitly if

$$3\alpha\beta = \alpha^2 - 16 \qquad (5.57a)$$

in which case the solution is

$$\Phi'^{(3)} = \tfrac{1}{3}(1 - \beta^2)\eta^3 + 3\xi^2\eta. \qquad (5.57b)$$

Finally, (5.51d) can be integrated if

$$3\alpha\beta = \beta^2 - 16 \qquad (5.58a)$$

leading to

$$\Phi^{(3)} = \tfrac{1}{6}(\alpha^2 + 2)\xi^3 - (\tfrac{1}{2}\alpha\beta + 5)\xi\eta^2. \qquad (5.58b)$$

However, the only common solution to (5.55a), (5.56), (5.57a) and (5.58a) is $\alpha = -2$, $\beta = 2$, leading to the PIW solutions, as we shall see in the next section. The situation is essentially not altered by considering linear relations of the form $\tau = \alpha\xi + \alpha'\eta$, $\sigma = \beta\eta + \beta'\xi$, where α', β' are constants. In this case we get simply 'duality' rotation, which reflects the fact that the field equations are invariant under a rotation in the (Φ, Φ') plane. This is most easily seen from (5.13a, b) which are invariant under $F \to e^{i\alpha}F$ for any real α.

5.7. Reduction to known solutions

In this section we carry out in detail the reduction of the approximate solution to the Bonnor and PIW solutions respectively to order λ^3. The reader who is not interested in these details may omit this section. We shall not discuss the reduction to Weyl solutions except to note that it can be obtained by taking τ to be proportional to ξ and setting $\sigma = \eta = 0$.

We first write down the Bonnor and PIW solutions in the forms which will be suitable for our purpose. The Bonnor solutions can be written as follows (the α and β are distinct to those in the last section):

$$f = h^{-1} = \alpha(1 + u^2)^{-1}[(u - B)^2 + A^2], \qquad (5.59a)$$

$$\Phi = \beta(1 + u^2)^{-1}u + \Phi_0, \quad \Phi' = -\tfrac{1}{2}\beta(1 + u^2)^{-1}(1 - u^2) + \Phi_0', \qquad (5.59b)$$

$$w_\rho = (2\lambda/\alpha A)\rho\sigma_z', \quad w_z = (2\lambda/\alpha A)\rho\sigma_\rho', \qquad (5.59c)$$

where α, β, A, B, Φ_0, Φ_0' are arbitrary constants and $\beta^2 = \alpha(1 + A^2 + B^2)$. The function u is given in terms of the harmonic function σ' by

$$u = A \tan \lambda \sigma' + B. \tag{5.59d}$$

This form of Bonnor's solution is slightly different from that given in his paper (1973). We have considered the case where his constants a and b satisfy $a^2 + b^2 < 4$, the opposite case being similar. When $a^2 + b^2 = 4$, Bonnor's solution becomes a special case of the PIW solutions. This fact is not apparent from (5.59a–d) but is not important for the present. We have also written $\lambda \sigma'$ for σ', where λ is a constant parameter, to facilitate comparison with the approximate solution.

The desired form of the PIW solution is as follows:

$$f^{-1} = h = (\lambda \xi' + 1)^2 + \lambda^2 \eta'^2, \tag{5.60a}$$

$$\Phi = (\lambda \xi' + 1)/[(\lambda \xi' + 1)^2 + \lambda^2 \eta'^2], \quad \Phi' = -\lambda \eta'/[(\lambda \xi' + 1)^2 + \lambda^2 \eta'^2], \tag{5.60b}$$

$$w_\rho = -2\rho[\lambda \eta'_z(\lambda \xi' + 1) - \lambda^2 \eta' \xi'_z], \quad w_z = 2\rho[\lambda \eta'_\rho(\lambda \xi' + 1) - \lambda^2 \eta' \xi'_\rho]. \tag{5.60c}$$

The form (5.60a, b, c) is obtained from the form (5.25) and (5.26) of the PIW solutions by setting $\eta = \lambda \xi' + 1$ and $\xi = \lambda \eta'$ in the latter, which is permissible since $\lambda \xi' + 1$ and $\lambda \eta'$ are harmonic, assuming that ξ', η' are harmonic functions.

By imposing suitable linear relations on the harmonic functions τ, σ, ξ, η one can get from (5.45), (5.46), (5.49a, b, c), (5.50) and (5.51a–d) the correct functions for the Weyl, Bonnor and PIW solutions up to and including coefficients of λ^3 in the power series in λ. For Bonnor's solution, let $\alpha A^2 = 1 + B^2$ in (5.59a, b, c). This ensures that h tends to unity at infinity if the harmonic function σ' is chosen to vanish at infinity. This also ensures that $h^{(0)} = 1$. Expanding Bonnor's solution (5.59a, b, c) in powers of λ, we get the following expansion coefficients (see (5.43)):

$$h^{(0)} = \frac{1 + B^2}{\alpha A^2} = 1, \quad h^{(1)} = \frac{2AB\sigma'}{1 + B^2},$$

$$h^{(2)} = \frac{(A^2 - 1 - B^2)}{1 + B^2}\sigma'^2, \quad h^{(3)} = -\frac{4AB\sigma'^3}{3(1 + B^2)}, \tag{5.61a}$$

$$w_\rho^{(1)} = (2/\alpha A)\rho\sigma'_z, \quad w_z^{(1)} = -(2/\alpha A)\rho\sigma'_\rho, \quad w^{(n)} = 0, \quad n \geqslant 0, \tag{5.61b}$$

$$\Phi^{(0)} = \frac{\beta B}{1 + B^2} + \Phi_0, \quad \Phi^{(1)} = \frac{\beta A(1 - B^2)}{(1 + B^2)^2}\sigma',$$

$$\Phi^{(2)} = \frac{\beta A^2 B(B^2 - 3)}{(1 + B^2)^3}\sigma'^2,$$

$$\Phi^{(3)} = \frac{\beta A}{3(1 + B^2)^4}[(1 + B^2)(1 - B^4) + 3A^2(-1 + 6B^2 - B^4)]\sigma'^3, \tag{5.61c}$$

$$\Phi'^{(0)} = -\frac{\beta(1 - B^2)}{2(1 + B^2)} + \Phi'_0, \quad \Phi'^{(1)} = \frac{2\beta AB\sigma'}{(1 + B^2)^2},$$

$$\Phi'^{(2)} = -\frac{\beta A^2(3B^2 - 1)}{(1 + B^2)^3}\sigma'^2$$

$$\Phi'^{(3)} = \frac{2\beta AB}{3(1 + B^2)^4}[(1 + B^2)^2 + 6A^2(B^2 - 1)]\sigma'^3. \tag{5.61d}$$

Let

$$(\tau, \sigma, \xi, \eta) = (a_1\sigma', a_2\sigma', a_3\sigma', a_4\sigma'), \tag{5.62a}$$

where a_i are constants given by

$$a_1 = \frac{2AB}{1 + B^2}, \quad a_2 = \frac{2A}{1 + B^2}, \quad a_3 = \frac{\beta A(1 - B^2)}{(1 + B^2)^2}, \quad a_4 = \frac{2\beta AB}{(1 + B^2)^2}. \tag{5.62b}$$

From (5.49a, b, c) we get

$$h^{(2)} = (\tfrac{1}{2}a_1^2 + \tfrac{1}{2}a_2^2 - a_3^2 - a_4^2)\sigma'^2,$$

$$\Phi'^{(2)} = \tfrac{1}{2}(-a_1a_4 + a_2a_3)\sigma'^2,$$

$$\Phi^{(2)} = \tfrac{1}{2}(-a_2a_4 - a_1a_3)\sigma'^2. \tag{5.63}$$

Substitution into (5.63) for the a_i from (5.62b) yields the same expression for $h^{(2)}, \Phi'^{(2)}, \Phi^{(2)}$ as in (5.61a, c, d), the expressions for $h^{(1)}, w^{(1)}, \Phi^{(1)}, \Phi'^1$ in (5.61a–d) being already consistent with (5.45), (5.46) and (5.62a, b). Further, (5.50) is consistent with $w^{(2)} = 0$, as in (5.61b). For the next order, consider (5.51a) which becomes

$$\nabla^2 h^{(3)} = a_1(a_1^2 + a_2^2 - 4a_3^2 - 4a_4^2)\sigma'(\sigma_\rho'^2 + \rho_z'^2), \tag{5.64a}$$

which has the solution

$$h^{(3)} = \tfrac{1}{6}a_1(a_1^2 + a_2^2 - 4a_3^2 - 4a_4^2)\sigma'^3. \tag{5.64b}$$

Similarly one gets

$$\Phi^{(3)} = \tfrac{1}{6}[a_3(a_1^2 - 2a_2^2 + 2a_3^2 + 2a_4^2) + 3a_1a_2a_4]\sigma'^3, \tag{5.65a}$$

$$\Phi'^{(3)} = \tfrac{1}{6}[a_4(a_1^2 - 2a_2^2 + 2a_3^2 + 2a_4^2) - 3a_1a_2a_3]\sigma'^3. \tag{5.65b}$$

Substitution for the a_i from (5.62b) yields, after some reduction, the same expressions for $h^{(3)}, \Phi'^{(3)}, \Phi^{(3)}$ as in (5.61a, c, d). Further (5.51b) is consistent with $w^{(3)} = 0$, as the right hand side of this equation vanishes for τ, σ, ξ, η given by (5.62a). Thus the series expansion reduces to the Bonnor solution, up to and including the coefficient of λ^3, for suitable choice of τ, σ, ξ, η.

For the PIW solution one gets, from (5.60a, b, c), the following expansion coefficients in powers of λ:

$$h^{(0)} = 1, \quad h^{(1)} = 2\xi', \quad h^{(2)} = \xi'^2 + \eta'^2, \quad h^{(n)} = 0, \quad n \geqslant 3, \qquad (5.66a)$$

$$w_\rho^{(1)} = -2\rho\eta_z', w_z^{(1)} = 2\rho\eta_\rho', \quad w_\rho^{(2)} = 2\rho(\eta'\xi_z' - \xi'\eta_z'),$$

$$w_z^{(2)} = 2\rho(\xi'\eta_\rho' - \eta'\xi_\rho'), \quad w^{(n)} = 0, \quad n \geqslant 3, \qquad (5.66b)$$

$$\Phi^{(0)} = 1, \quad \Phi^{(1)} = -\xi', \quad \Phi^{(2)} = \xi'^2 - \eta'^2, \quad \Phi^{(3)} = \xi'(-\xi'^2 + 3\eta'^2),$$

$$(5.66c)$$

$$\Phi'^{(0)} = 0, \quad \Phi'^{(1)} = -\eta', \quad \Phi'^{(2)} = 2\xi'\eta', \quad \Phi'^{(3)} = -\eta'(3\xi'^2 - \eta'^2).$$

$$(5.66d)$$

For this solution we choose

$$(\tau, \sigma, \xi, \eta) = (2\xi', -2\eta', -\xi', -\eta'). \qquad (5.67)$$

This is consistent with the expressions for $h^{(1)}, w^{(1)}, \Phi^{(1)}, \Phi'^{(1)}$ in (5.45), (5.46) and (5.66a–d). It is also readily verified that with this choice (5.49a, b, c) and (5.50) yield the same $h^{(2)}, w^{(2)}, \Phi^{(2)}, \Phi'^{(2)}$ as in (5.66a–d). The right hand sides of (5.51a, b) vanish, so that these equations are consistent with $h^{(3)} = w^{(3)} = 0$. Equations (5.51c, d) become respectively

$$\nabla^2\Phi'^{(3)} = 6\eta'(\eta_\rho'^2 + \eta_z'^2) - 6\eta'(\xi_\rho'^2 + \xi_z'^2) - 12\xi'(\xi_\rho'\eta_\rho' + \xi_z'\eta_z'), \qquad (5.68a)$$

$$\nabla^2\Phi^{(3)} = -6\xi'(\xi_\rho'^2 + \xi_z^2) + 6\xi'(\eta_\rho'^2 + \eta_z'^2) + 12\eta'(\xi_\rho'\eta_\rho' + \xi_z'\eta_z'^2) \qquad (5.68b)$$

It is readily verified, with the use of (5.54), that $\Phi'^{(3)}$ and $\Phi^{(3)}$ given by (5.66c, d) satisfy (5.68a, b). The series expansion is thus also consistent with the PIW solution.

5.8. Physical interpretation

Since space–time becomes flat and the fields vanish when $\lambda = 0$, the approximate solution represents a weak field in some sense. For example, the constant λ can be taken to be proportional both to the gravitational constant and the charge of the sources so that when λ is zero the gravitational constant vanishes so that space is flat and the charges giving rise to the electromagnetic field vanish so that the fields are zero. For convenience we consider the functions up to λ^2 terms only. To examine the possible sources for these solutions, we choose the harmonic functions so that space–time is asymptotically flat to this order. To this end let

$$\tau = \tau' + k_1\xi + k_2\eta, \quad k_1, k_2 \text{ constant} \qquad (5.69)$$

where τ' is a harmonic function, and let τ, σ, ξ, η have the following forms at infinity:

$$\left.\begin{array}{l} \tau' = A_1 r^{-1} + A_1' zr^{-3} + \cdots, \quad \sigma = A_2 zr^{-3} + A_2'(r^{-3} - 3z^2r^{-5}) + \cdots \\ \xi = B_1 r^{-1} + B_1' zr^{-3} + \cdots, \quad \eta = B_2 r^{-1} + B_2' zr^{-3} + \cdots \end{array}\right\} (5.70)$$

where $r^2 = \rho^2 + z^2$, and the dots represent terms which vanish at infinity faster than the preceding terms. The reason for choosing the leading term for σ as in (5.70) is that this results in the correct asymptotic behaviour for w (see (2.35) and (2.36)), which must tend to zero at infinity like $\rho^2 r^{-3}$. The behaviour of the functions up to λ^2 terms and up to r^{-2} terms is then as follows:

$$h = 1 + \lambda[(A_1 + k_1 B_1 + k_2 B_2)r^{-1} + (A_1' + k_1 B_1' + k_2 B_2')zr^{-3}]$$
$$+ \lambda^2[\tfrac{1}{2}(A_1 + k_1 B_1 + k_2 B_2)^2 - B_1^2 - B_2^2]r^{-2}, \tag{5.71a}$$
$$w = -\lambda(A_2 \rho^2 r^{-3} - 3A_2' \rho^2 z r^{-5}) + \lambda^2 A' \rho^2 r^{-4},$$
$$A' \equiv -A_2(A_1 + k_1 B_1 + k_2 B_2) + 2(B_2 B_1' - B_2' B_1), \tag{5.71b}$$
$$\Phi = \lambda(B_1 r^{-1} + B_1' z r^{-3}) - \lambda^2 B_1(A_1 + k_1 B_1 + k_2 B_2)r^{-2}, \tag{5.71c}$$
$$\Phi' = \lambda(B_2 r^{-1} + B_2' z r^{-3}) - \lambda^2 B_2(A_1 + k_1 B_1 + k_2 B_2)r^{-2}. \tag{5.71d}$$

These equations represent fairly general asymptotic behaviour. From (2.36) and (5.71a) we see that the expression $\tfrac{1}{2}\lambda(A_1 + k_1 B_1 + k_2 B_2)$ is proportional to the mass of the source. Thus we can have the correct asymptotic behaviour for w given by (5.71b) and yet have a non-zero mass, unlike the Bonnor or Papapetrou solutions. For the Bonnor solution σ is proportional to τ (see (5.62a)) which means that if σ is to have the asymptotic behaviour as in (5.70), from (5.69) and (5.70) one must have

$$A_1 + k_1 B_1 + k_2 B_2 = 0 \tag{5.72}$$

so that the mass of the source is zero. Similarly, one can verify that the conditions (5.67) for the PIW solutions imply the following relations:

$$A_1 + k_1 B_1 + k_2 B_2 = -2B_1, \quad B_2 = 0, \quad 2B_2' = A_2. \tag{5.73}$$

These relations imply that the corresponding sources satisfy (5.1), since, for example, $\lambda B_2'$ gives the magnetic moment (being the coefficient of zr^{-3} in the magnetic potential Φ' (5.71d)) and $\tfrac{1}{2}\lambda A_2$ is the angular momentum (compare (5.71b) with (2.35) and (2.36)). The first equation in (5.73) shows that the mass is equal to the charge since $\tfrac{1}{2}\lambda(A_1 + k_1 B_1 + k_2 B_2)$ is the mass (see (5.71a) and (2.36)) and λB_1 is the charge (this being the coefficient of the r^{-1} term in Φ (5.71c). With the use of the techniques of this section one can also show that the sources for the class S of solutions satisfy (5.2). In the absence of magnetic monopoles, we should have $B_2 = 0$.

This completes our detailed analysis of the class of approximate solutions of the EM equations depending on four harmonic functions. We recapitulate the reason for considering this problem in detail. Firstly, this is an example of a problem in the general area of the subject matter of this book carried out to a certain stage of completion. Secondly, it gives an insight into the structure of the general solution of the axisymmetric stationary EM exterior equations and the manner in which this general solution is related to the sources that give rise to the solution. A number of further questions arise with regard to this approximation scheme. It has already been remarked that the Herlt (1978) class of solutions does not fit into this scheme and it is interesting to consider the modification which will include the Herlt solutions. Because of this fact it is clear that the above approximate solution is not necessarily the most general approximate solution. We also saw that at every stage of the approximation scheme one can add an arbitrary harmonic function to the various functions. It is interesting to examine the precise nature of the dependence of the general solution on these additional harmonic functions.

5.9. Further remarks about the Einstein–Maxwell equations

We saw in Section 5.3 how to establish a relation between solutions of the pure Einstein equations and certain solutions of the EM equations. Starting with the Kerr solution, it is possible to use this correspondence to find a solution of the EM equations leading to the following metric (Ernst, 1968b):

$$ds^2 = [1 - (2mr - e^2)\Sigma_1^{-1}] dt^2 - \sin^2 \theta [r^2 + a^2$$
$$+ a^2 \sin^2 \theta (2mr - e^2)\Sigma_1^{-1}] d\phi^2$$
$$- 2a \sin^2 \theta (2mr - e^2)\Sigma_1^{-1} d\phi\, dt - \Sigma_1(\Sigma_2^{-1} dr^2 + d\theta^2), \quad (5.74)$$

with

$$\Sigma_1 \equiv r^2 + a^2 \cos^2 \theta, \quad \Sigma_2 \equiv r^2 + a^2 + e^2 - 2mr.$$

This metric reduces to the Kerr metric (3.27) when $e = 0$. This metric was discovered first by Newman *et al.* (1965), and represents the exterior field of a rotating charged black hole with charge e and is known as the Kerr–Newman metric.

When $w = 0$, $\Phi = 0$, the source of the electromagnetic field is non-rotating and one gets the axially symmetric static Einstein–Maxwell equations. One can get the Weyl electrostatic solution from this by assuming f to be a function of Φ: $f = f(\Phi)$. Herlt (1978) found a class of solutions of these equations (this class was mentioned earlier) that is distinct from the Weyl class, which is given in terms of a harmonic function and which contains asymptotically flat solutions. For the electrostatic EM equations too one can establish a certain connection with the rotating solution of the pure Einstein equations. Through this correspondence Bonnor (1966) generated a solution of the electrostatic EM equations from the Kerr solution, the new solution representing the field of a magnetic dipole. This Bonnor solution is in fact a member of the Herlt class.

5.10. Generation of solutions

A number of authors have considered using known solutions of the Einstein and EM equations to generate new solutions. Kinnersley (1977) and Kinnersley and Chitre (1977, 1978a, b) have used group-theoretic methods following earlier work of Geroch (1972) to generate asymptotically flat solutions from known ones. We have already encountered at the end of Section 5.6 a trivial example of the invariance of the EM equations under a group of transformations, namely, the invariance of (5.13a, b) under $F \rightarrow$ $e^{i\alpha} F$ where α is a real constant. This group is the group under ordinary multiplication of complex unimodular numbers, that is, numbers of the form $e^{i\alpha}$ where α is real, and is denoted by $U(1)$. One can find more complicated non-trivial groups of transformation under which the EM equations are invariant and this enables one to generate new solutions by applying these transformations to known solutions. (See also Hoenselaers (1979), Hoenselaers, Kinnersley and Xanthopoulos (1979).)

Harrison (1978), Neugebauer (1979) and Belinsky and Zakharov (1978, 1979) have reformulated the stationary axisymmetric Einstein equations using Bäcklund transformations and soliton theory. Belinsky and Zakharov and Hauser and Ernst (1979, 1980) have also formulated the problem in terms of linear integral equations connected with the inverse scattering and homogeneous Hilbert problem. The relation between these approaches has been considered by Cosgrove (1980, 1982). One of the central problems, namely the finding of all asymptotically flat solutions, is still unsolved but the developments mentioned briefly in this section appear to be promising.

6

Rotating charged dust

6.1. Introduction

In this chapter we shall deal with the problem of charged dust (pressureless matter) rotating steadily about an axis of symmetry. The rotation is steady in the sense that the motion is stationary, i.e. independent of time. The dust distribution has axial symmetry. The forces acting on a typical portion of the dust are centrifugal, gravitational, electric and magnetic.

The problem of rotating charged dust might not be astrophysically interesting, but it is nevertheless important because it pertains to a well defined physical situation in which the interplay of several forces can be studied, both in Newton–Maxwell theory and in general relativity, and a fruitful comparison can be made. As mentioned earlier, it is rare to find interior solutions for the Einstein or the Einstein–Maxwell equations. This problem has already yielded a number of exact interior solutions of the Einstein–Maxwell equations for a physically well defined energy-momentum tensor. A great deal remains to be understoood about this problem and I believe that when a complete analysis of it is carried out much useful insight into the physical content of general relativity will be gained.

This chapter is based almost entirely on the author's work (Islam, 1977, 1978*a*, 1979, 1980, 1983*a*, *b*, *c*, 1984; see also Boachie and Islam, (1983), and Islam, Van den Bergh and Wils, 1984). After getting an introduction to this problem here the interested reader can refer to other work. Among earlier papers on charged dust are those of Som and Raychaudhuri (1968), De and Raychaudhuri (1968), Banerjee and Banerji (1968), Raychaudhuri and De (1970), Banerjee, Chakravarty and Dutta Choudhury (1976), Klostar and Das (1977). Among more recent papers are those of Bonnor (1980*b*) Raychaudhuri (1982) and Van den Burgh and Wils (1984*a*, *b*, *c*, *d*).

6.2. Newton–Maxwell theory

Let the number density of particles be n, and the mass and the charge of the particles be m and q respectively. Throughout this chapter we shall confine ourselves to the case where the ratio of the mass and charge density of the particles is constant, that is, independent of position (see Bonnor (1980b) for the other case). Thus m and q are constant, mn being the mass density and qn the charge density; n is in general a function of position. There is no pressure. We envisage an axially symmetric situation in which the particles are rotating about an axis of symmetry (the z-axis) with an angular velocity Ω which is a function of ρ (the distance from the axis) and z only (see Fig. 1.1). We ask if a stationary (that is, independent of time) equilibrium is possible. As usual we set the gravitational constant and velocity of light equal to unity. The equation of motion is as follows:

$$mn(\partial/\partial t + \mathbf{v}\cdot\nabla\mathbf{v})\mathbf{v} = \mathbf{J} \wedge \mathbf{B} - mn\nabla V + qn\mathbf{E}, \tag{6.1}$$

where $\mathbf{v}$ is the velocity of the particles, constrained to be given by the following cartesian components:

$$\mathbf{v} = (-\Omega y, \Omega x, 0), \quad \Omega = \Omega(\rho, z). \tag{6.2}$$

$\mathbf{J}$ is the current density given by

$$\mathbf{J} = qn\mathbf{v}, \tag{6.3}$$

while $\mathbf{E}$ and $\mathbf{B}$ are the electric field and the magnetic induction field respectively, satisfying (in the stationary situation)

$$\text{curl } \mathbf{E} = 0, \quad \text{div } \mathbf{E} = 4\pi qn, \tag{6.4}$$

$$\text{div } \mathbf{B} = 0, \quad \text{curl } \mathbf{B} = 4\pi\mathbf{J}. \tag{6.5}$$

V is the Newtonian gravitation potential satisfying

$$\nabla^2 V = 4\pi mn. \tag{6.6}$$

From (6.4) we have

$$\mathbf{E} = \nabla\Phi, \quad \nabla^2\Phi = 4\pi qn, \tag{6.7}$$

where Φ is the electrostatic potential. Because of the symmetry of the situation, V and Φ are functions of ρ and z only. In cartesian components, with $\mathbf{B} = (B^{(1)}, B^{(2)}, B^{(3)})$, (6.1) reads ($\partial\mathbf{v}/\partial t = 0$ because of stationarity):

$$m\Omega^2(-x, y, 0) = q\Omega(xB^{(3)}, yB^{(3)}, -xB^{(1)} - yB^{(2)}) - m(V_x, V_y, V_z)$$
$$+ q(\Phi_x, \Phi_y, \Phi_z), \tag{6.8}$$

where $V_x \equiv \partial V/\partial x$, etc. Equation (6.8) is only valid under the assumption $n \neq 0$ (i.e. inside matter) since a factor n has been cancelled. For any axially symmetric function $f(\rho, z)$ we have $f_x = x\rho^{-1}f_\rho$, $f_y = y\rho^{-1}f_\rho$. The second equation in (6.5) reads

$$(B_y^{(3)} - B_z^{(2)}, B_z^{(1)} - B_x^{(3)}, \quad B_x^{(2)} - B_y^{(1)}) = 4\pi\Omega qn(-y, x, 0). \qquad (6.9)$$

The current has to satisfy the equation of continuity

$$\text{div } \mathbf{J} = q \, \text{div} \, (n\mathbf{v}) = 0. \qquad (6.10)$$

For $\mathbf{v}$ given by (6.2) this is satisfied identically if n is a function of ρ and z only: $n = n(\rho, z)$.

From the third component of (6.8) we see that $xB^{(1)} + yB^{(2)}$ must be a function of ρ and z. This implies the following:

$$B^{(1)} = xB + yC, \quad B^{(2)} = yB - xC, \qquad (6.11)$$

where B and C are functions of ρ and z only and correspond respectively to the radial and azimuthal components of the magnetic induction.

Consider the integral of the magnetic induction around a circle C' given by $z = \text{constant}$, $\rho = \text{constant}$. This line integral is proportional to the current flowing through C' and must vanish identically for dust rotating around the z-axis. With the use of (6.11) we have

$$\oint_{C'} \mathbf{B} \cdot d\mathbf{s} = - \int \rho^2 C \, d\phi = 0, \qquad (6.12)$$

where ϕ is the azimuthal angle. Thus we must have $C = 0$. The first equation in (6.5) together with (6.11) and $C = 0$ implies

$$\rho^2 B = b_z, \quad \rho B^{(3)} = -b_\rho, \qquad (6.13)$$

where b is an arbitrary function of ρ and z. With the use of (6.13) the three equations in (6.8) can be reduced to the following two:

$$-q\Omega b_z - mV_z + q\Phi_z = 0, \qquad (6.14)$$

$$q\Omega b_\rho + mV_\rho - q\Phi_\rho = m\Omega^2 \rho. \qquad (6.15)$$

The equations so far are valid for differential rotation when Ω is a function of ρ and z. We now assume that Ω is a constant, that is, we have rigid rotation. Equation (6.14) can then be integrated to yield

$$-mV + q\Phi - q\Omega b = \alpha(\rho) \qquad (6.16)$$

where $\alpha(\rho)$ is an arbitrary function of ρ. Equations (6.15) and (6.16) imply

$$\alpha(\rho) = -\tfrac{1}{2}\Omega^2 \rho^2 + \alpha_0 \qquad (6.17)$$

where α_0 is an arbitrary constant. The third component of (6.9) is satisfied identically, while the first two components yield the single equation

$$\Delta b = b_{\rho\rho} + b_{zz} - \rho^{-1} b_\rho = 4\pi q\Omega\rho^2 n. \qquad (6.18)$$

By applying the

$$\nabla^2 \equiv \partial^2/\partial\rho^2 + \partial^2/\partial z^2 + \rho^{-1}\partial/\partial\rho$$

operator to (6.16) and using (6.6), (6.7) and (6.18) we get the following relation between n and b:

$$2\pi(m^2 - q^2 + q^2\Omega^2\rho^2)n = -q\Omega\rho^{-1}b_\rho + m\Omega^2. \tag{6.19}$$

Eliminating n between (6.18) and (6.19) we get the following equation for b:

$$b_{\rho\rho} + b_{zz} + [-\rho^{-1} + 2q^2\Omega^2\rho(m^2 - q^2 + q^2\Omega^2\rho^2)^{-1}]b_\rho$$
$$= 2mq\Omega^3\rho^2(m^2 - q^2 + q^2\Omega^2\rho^2)^{-1}. \tag{6.20}$$

This can be regarded as the fundamental equation for rigidly rotating charged dust in Newton–Maxwell theory. Once this equation is solved the other functions can be obtained from the solution. The number density n can be obtained from (6.19) and the magnetic induction from (6.13). The gravitational and electric potentials V and Φ can be obtained, in principle, from the Poisson equations (6.6) and (6.7).

As a limiting case one can set $b_z = 0$, leading to $B^{(1)} = B^{(2)} = 0$ and a cylindrically symmetric configuration. The resulting equation (6.20) is then an ordinary differential equation which can be solved explicitly and from which the other functions can be obtained. The solution is given as follows:

$$B^{(3)} = (q\Omega)^{-1}(a_0 - m\Omega^2\rho^2)(a + \rho)^{-1}, \tag{6.21a}$$

$$n = a'(a + \rho^2)^{-2} \tag{6.21b}$$

$$V = m\pi a'a^{-1}\log(a + \rho^2) + V_0, \tag{6.21c}$$

$$\Phi = q\pi a'a^{-1}\log(a + \rho^2) + \Phi_0, \tag{6.21d}$$

where

$$a \equiv (q^2\Omega^2)^{-1}(m^2 - q^2), \quad a' = (2\pi q^2\Omega^2)^{-1}[mq^{-2}(m^2 - q^2) + a_0] \tag{6.21e}$$

and a_0, Φ_0, V_0 are arbitrary constants. The solution given by (6.21a–e) is well behaved everywhere for $q^2 < m^2$, with n positive, for a suitable choice of a_0. Let there be a boundary of the cylinder of dust at $\rho = \rho_0$. Then the number of particles per unit z-coordinate is

$$N_0 = \int_0^{\rho_0} 2\pi\rho n \, d\rho = \pi a'\rho_0^2(a + \rho_0^2)^{-1}, \tag{6.22}$$

which is positive and well defined for $a' > 0$ and $m^2 > q^2$. Note that N_0 remains finite even when ρ_0 tends to infinity.

Consider now an exterior solution to (6.21a–e), which can be shown to be given by the following:

$$V = V_0'\log\rho + V_1, \tag{6.23a}$$

$$\Phi = \Phi_0'\log\rho + \Phi_1, \tag{6.23b}$$

$$B^{(3)} = B_0, \tag{6.23c}$$

where V'_0, Φ'_0, B'_0, V_1, Φ_1 are constants. In (6.21a–e) and (6.23a,b,c) the function $B^{(3)}$ must be continuous and Φ, V and their derivatives must be continuous at $\rho = \rho_0$. This can be achieved by the following choice for the various constants:

$$V'_0 \log \rho_0 + V_1 = \pi m a' a^{-1} \log (a + \rho_0^2) + V_0, \tag{6.24a}$$

$$\Phi'_0 \log \rho_0 + \Phi_1 = \pi q a' a^{-1} \log (a + \rho_0^2) + \Phi_0, \tag{6.24b}$$

$$V'_0 = 2\pi m a' a^{-1} \rho_0^2 (a + \rho_0^2)^{-1} = m q^{-1} \Phi'_0, \tag{6.24c}$$

$$B_0 = (q\Omega)^{-1} (a_0 - m\Omega^2 \rho_0^2)(a + \rho_0^2)^{-1}. \tag{6.24d}$$

Thus for $q^2 < m^2$ (6.21a–e), (6.23a,b,c) and (6.24a–d) define a global solution (the interior solution matched smoothly to the exterior) which is regular everywhere. Such global solutions in Newton–Maxwell theory are useful to have as models for which one can try to find the analogue in general relativity. We note from (6.23c) and (6.24d) that the magnetic field outside the charged dust cylinder is constant and along the z-axis, its direction being dependent on the sense of rotation, that is, the sign of Ω as in (6.24d).

The solution (6.21a–e) displays singularities for $q^2 > m^2$ (since then a is negative and so, for example, n tends to infinity for $\rho^2 + a = 0$). This probably reflects the fact that for $q^2 > m^2$ the electrostatic forces overwhelm the gravitational, centrifugal and magnetic forces and we cannot have equilibrium configurations. We shall find a similar situation in general relativity.

Reverting to axial symmetry, when $m = q$, (6.20) reduces to

$$\nabla^2 b = 2\Omega \tag{6.25}$$

which has the solution

$$b = \eta + \tfrac{1}{2}\Omega \rho^2 \tag{6.26}$$

where η is a harmonic function:

$$\nabla^2 \eta = 0. \tag{6.27}$$

The solution (6.26), however, does not lead to an explicit form for the other functions. For example (6.6) with the use of (6.19) with $m = q$ reduces to

$$\nabla^2 V = -2(\Omega \rho^3)^{-1} \eta_\rho, \tag{6.28}$$

which is difficult to solve explicitly. One can, however, find an explicit particular solution in this case (Islam, 1983a).

6.3. Solution for differential rotation in Newton–Maxwell theory

We consider differential rotation in this section where Ω is again a function of ρ and z: $\Omega = \Omega(\rho, z)$. Equations (6.15) and earlier equations are valid in this case. Differentiating (6.15) with respect to z and (6.14) with respect to ρ and adding, we get

$$q(\Omega_z b_\rho - \Omega_\rho b_z) = 2m\rho\Omega\Omega_\rho. \tag{6.29}$$

Again the third component of (6.9) is satisfied identically while the first two components yield the equation

$$\Delta b = 4\pi q\rho^2 \Omega n. \tag{6.30}$$

From (6.14) and (6.15) we also get

$$m\nabla^2 V - q\nabla^2 \Phi = -q\Omega\nabla^2 b - q(\Omega_\rho b_\rho + \Omega_z b_z) + 2m\Omega^2 + 2m\rho\Omega\Omega_\rho \tag{6.31}$$

We substitute for $\nabla^2 V$ and $\nabla^2 \Phi$ from (6.6) and (6.7) respectively and for $b_{\rho\rho} + b_{zz}$ from (6.30) to get the following equation:

$$4\pi(m^2 - q^2 + q^2\Omega^2\rho^2)n = -qb_\rho(2\rho^{-1}\Omega + \Omega_\rho) - qb_z\Omega_z + 2m\Omega^2 + 2m\rho\Omega\Omega_\rho. \tag{6.32}$$

To solve (6.29), we consider b as a function of ρ and Ω:

$$b = b(\rho, \Omega), \tag{6.33a}$$

and let

$$b' = \left(\frac{\partial b}{\partial \Omega}\right)_{\rho = \text{constant}}, \quad \dot{b} = \left(\frac{\partial b}{\partial \rho}\right)_{\Omega = \text{constant}} \tag{6.33b}$$

so that

$$b_\rho = b'\Omega_\rho + \dot{b}, \quad b_z = b'\Omega_z. \tag{6.34}$$

Substituting from (6.34) into (6.29) we get

$$q\dot{b} = 2\rho\Omega, \tag{6.35}$$

which has the solution

$$qb = \rho^2\Omega + \omega(\Omega), \tag{6.36}$$

where $\omega(\Omega)$ is an arbitrary function of Ω. We now eliminate b and n from (6.32) with the use of (6.30) and (6.36) to get the following single equation for Ω:

$$\begin{aligned}(m\rho^2 + \omega')\Delta\Omega &+ \omega''(\Omega_\rho^2 + \Omega_z^2) + 4m\rho\Omega_\rho \\ &+ q^2\rho^2\Omega(m^2 - q^2 + q^2\rho^2\Omega^2)^{-1}[(m\rho^2 + \omega')(\Omega_\rho^2 + \Omega_z^2) \\ &+ 2m\Omega^2 + 2m\rho\Omega\Omega_\rho + 2\rho^{-1}\omega'\Omega\Omega_\rho] = 0,\end{aligned} \tag{6.37}$$

where $\Omega' \equiv \partial\omega/\partial\Omega$, etc. Equation (6.37) is the fundamental equation for differential rotation in Newton–Maxwell theory. This equation becomes determinate once an arbitrary function of one variable, namely $\omega(\Omega)$, is specified. The general relativistic problem of differential rotation can be brought to a similar state in the sense that the problem can be reduced to a system of three coupled equations which become determinate once an arbitrary function of Ω is specified (Islam, 1979; see also Section 6.9).

We have so far not succeeded in solving (6.37). However, a general solution of this equation can be found when the mass equals the charge. Setting $m = q$ in (6.37) we get the following equation:

$$\Omega(m\rho^2 + \omega')\nabla^2\Omega + (m\rho^2 + \omega' + \Omega\omega'')(\Omega_\rho^2 + \Omega_z^2)$$
$$+ 4m\rho\Omega\Omega_\rho + 2m\Omega^2 = 0. \tag{6.38}$$

We seek a solution of this equation in terms of a harmonic function η satisfying (6.27). Assume η to be a function of ρ and Ω:

$$\eta = \eta(\rho, \Omega), \tag{6.39}$$

and define η' and $\dot\eta$ as in (6.33b). Then

$$\eta_\rho = \dot\eta + \eta'\Omega_\rho, \tag{6.40a}$$
$$\eta_{\rho\rho} = \ddot\eta + 2\dot\eta\Omega_\rho + \eta''\Omega_\rho^2 + \eta'\Omega_{\rho\rho}, \tag{6.40b}$$
$$\eta_{zz} = \eta''\Omega_z^2 + \eta'\Omega_{zz}. \tag{6.40c}$$

The equation $\nabla^2\eta = 0$ can then be written as

$$\nabla^2\eta = \eta'\nabla^2\Omega + \eta''(\Omega_\rho^2 + \Omega_z^2) + \ddot\eta + \rho^{-1}\dot\eta + 2\dot\eta'\Omega_\rho = 0. \tag{6.41}$$

Comparing (6.41) with (6.38) we see that the two equations coincide if

$$\left.\begin{array}{l} \eta' = \Omega(m\rho^2 + \omega'), \quad \eta'' = m\rho^2 + \omega' + \Omega\omega'', \\ \dot\eta' = 2m\rho\Omega, \quad \ddot\eta + \rho^{-1}\dot\eta = 2m\Omega^2. \end{array}\right\} \tag{6.42}$$

It is readily verified that (6.42) is satisfied if

$$\eta = \tfrac{1}{2}m\rho^2\Omega^2 + H(\Omega), \tag{6.43}$$

where $H(\Omega)$ is a function of Ω only given by

$$H(\Omega) = \int^\Omega \Omega\omega' \, d\Omega. \tag{6.44}$$

For any given function $\omega(\Omega)$, the function $H(\Omega)$ can be evaluated from (6.44). Then Ω is given implicitly as a function of ρ and η by (6.43).

We now show that the choice $\omega' = 0$ leading to $H(\Omega) = 0$ gives an unphysical solution, for in this case we get from (6.36), (6.43) and (6.32) (with $m = q = 1$) the following expressions:

$$\Omega = \rho^{-1}(2\eta)^{1/2}, \tag{6.45a}$$
$$b = \rho(2\eta)^{1/2}, \tag{6.45b}$$
$$n = (4\pi)^{-1}[-\rho^{-2} - (4\eta^2)^{-1}(\eta_\rho^2 + \eta_z^2)]. \tag{6.45c}$$

From (6.45c) we see that the number density n is necessarily negative. Let us nevertheless determine the corresponding Φ and V. From (6.6), (6.7) and

(6.45c) we get

$$\nabla^2\Phi = \nabla^2 V = -\rho^{-2} - (4\eta^2)^{-1}(\eta_\rho^2 + \eta_z^2). \tag{6.46}$$

We seek a solution of (6.46) in the form

$$\Phi = \Phi^{(1)}(\rho) + \Phi^{(2)}(\eta), \quad V = V^{(1)}(\rho) + V^{(2)}(\eta), \tag{6.47}$$

where $\Phi^{(1)}$, $V^{(1)}$ are functions of ρ only and $\Phi^{(2)}$, $V^{(2)}$ are functions of η only. Substituting in (6.46) from (6.47) we get ordinary differential equations for $\Phi^{(1)}$, $V^{(1)}$, $\Phi^{(2)}$, $V^{(2)}$ which can be solved to yield the following solutions:

$$\Phi = \tfrac{1}{4}\log\eta + \Phi_1\eta - \tfrac{1}{2}(\log\rho)^2 + \Phi_0\log\rho + \Phi_0', \tag{6.48a}$$

$$V = \tfrac{1}{4}\log\eta + V_1\eta - \tfrac{1}{2}(\log\rho)^2 + V_0\log\rho + V_0'. \tag{6.48b}$$

where Φ_0, Φ_0', Φ_1, V_0, V_0', V_1 are arbitrary constants. With the use of (6.45a, b) and (6.48a, b) it can be verified that (6.14) and (6.15) are satisfied if $\Phi_0 = V_0$ and $1 + V_1 = \Phi_1$. Thus (6.45a, b, c) and (6.48a, b) constitute a class of exact solutions in terms of the harmonic function η. Such a complete solution is necessary if one is to look for the general relativistic analogues of solutions in Newton–Maxwell theory. Although this solution is unphysical because of the negative sign of n given by (6.45c), its existence indicates that in the corresponding case a solution may exist in general relativity in terms of a harmonic function. Whether the general relativistic solution, if it exists, will also be unphysical is an open question. Interior solutions in general relativity are rare and even an unphysical solution is of some interest. Besides, the possibility exists that an apparently unphysical solution could in future be given an alternative interesting interpretation.

We now consider a cylindrically symmetric solution with differential rotation and with $m = q = 1$. In this case (6.31) becomes

$$\Omega b_{\rho\rho} + (\rho^{-1}\Omega + \Omega_\rho)b_\rho = 2\Omega^2 + 2\rho\Omega\Omega_\rho, \tag{6.49}$$

which, after multiplication by ρ, can be interpreted as follows:

$$\rho\Omega b_\rho = \rho^2\Omega^2 + A, \tag{6.50}$$

where A is an arbitrary constant. The corresponding number density, from (6.32), is given by

$$n = (4\pi\rho\Omega)^{-1}\Omega_\rho - (4\pi)^{-1}A[2(\rho^4\Omega^2)^{-1} + (\rho^3\Omega^3)^{-1}\Omega_\rho]. \tag{6.51}$$

To find a regular solution, we choose $A = 0$. Then Φ and V satisfy the following equations:

$$\Phi_{\rho\rho} + \rho^{-1}\Phi_\rho = V_{\rho\rho} + \rho^{-1}V_\rho = (\rho\Omega)^{-1}\Omega_\rho, \tag{6.52}$$

which can be integrated as follows:

$$\rho V_\rho = \log\Omega + V_0, \tag{6.53a}$$

$$\rho\Phi_\rho = \log\Omega + \Phi_0, \qquad (6.53b)$$

where Φ_0, V_0 are arbitrary constants. Equation (6.15) is satisfied by $\Phi_0 = V_0$. So far the function $\Omega(\rho)$ is arbitrary. Choose $\Omega(\rho)$ as follows:

$$\Omega = a_0 + a_1\rho^2, \qquad (6.54)$$

where a_0 and a_1 are constants. Then (6.51) with $A = 0$ gives

$$n = (2\pi)^{-1}a_1(a_0 + a_1\rho^2)^{-1}, \qquad (6.55)$$

which is positive and well behaved everywhere if $a_0, a_1 > 0$. If $\rho = \rho_0$ is the boundary of the cylindrical charge distribution then the number per unit z-coordinate is

$$N_0 = \int_0^{\rho_0} 2\pi n\rho \, d\rho = \tfrac{1}{2}\log(1 + a_0^{-1}a_1\rho_0^2). \qquad (6.56)$$

The corresponding magnetic field is (see (6.23) and (6.50)).

$$B^{(1)} = B^{(2)} = 0, \quad B^{(3)} = -a_0 - a_1\rho^2. \qquad (6.57)$$

From (6.53a, b) we get

$$\Phi_\rho = V_\rho = \rho^{-1}\log(a_0 + a_1\rho^2) + \Phi_0\rho^{-1}, \qquad (6.58)$$

which, with the choice $\Phi_0 = 0$ is well behaved as ρ tends to zero if we choose $a_0 = 1$. It is adequate to determine the derivatives Φ_ρ, V_ρ of the potentials since these represent the forces.

The exterior field is again given by (6.23a, b, c). To match the above interior solution smoothly at $\rho = \rho_0$ we get from (6.23a, b, c), (6.57) and (6.58)

$$\Phi_0' = V_0' = \log(1 + a_1\rho_0^2), \quad B_0 = -1 - a_1\rho_0^2. \qquad (6.59)$$

Thus (6.23a, b, c), (6.54), (6.55), (6.57), (6.58) and (6.59) give a regular global solution, that is, an interior solution which matches smoothly on to an exterior solution and which is regular everywhere. It is useful to have such regular solutions so that one can look for the general relativistic analogue and make a comparison.

From (6.14) and (6.15) it is clear that the terms representing the Lorentz force on the particles are those given in terms of Φ and b. Thus if the Lorentz force vanishes, we must have

$$\Phi_z - \Omega b_z = 0, \quad V_z = 0, \qquad (6.60a)$$

$$\Phi_\rho - \Omega b_\rho = 0, \quad V_\rho = \Omega^2\rho. \qquad (6.60b)$$

Equations (6.60a, b) imply that one has either rigid rotation ($\Omega = $ constant) or cylindrical symmetry (Van den Bergh and Wils, 1984c).

This completes our analysis of the problem of charged rotating dust in Newtonian physics. It is useful to have studied this simpler situation before embarking on the more complicated study of the same problem in general relativity. Our main object in this chapter is to find a simple well defined physical situation in Newtonian physics and to try to find its analogue in general relativity. As we shall see, even this simple programme is not easy to complete and we shall have only limited success.

6.4. The Einstein–Maxwell interior equations

The Einstein–Maxwell interior equations are as follows:

$$R_{\mu\nu} - \tfrac{1}{2}g_{\mu\nu}R = 8\pi(T_{\mu\nu} + E_{\mu\nu}), \tag{6.61a}$$

$$F_{\mu\nu;\sigma} + F_{\nu\sigma;\mu} + F_{\sigma\mu;\nu} = 0, \tag{6.61b}$$

$$F^{\mu\nu}{}_{;\nu} = -4\pi J^{\mu}, \tag{6.61c}$$

where as usual a semicolon denotes covariant differentiation, $E_{\mu\nu}$ is the electromagnetic energy-momentum tensor given by (see (6.71) and (5.3a, b, c) for definition of $F_{\mu\nu}$)

$$E_{\mu\nu} = (4\pi)^{-1}(-F_{\mu\alpha}F^{\alpha}_{\nu} + \tfrac{1}{4}g_{\mu\nu}F_{\alpha\beta}F^{\alpha\beta}). \tag{6.62}$$

Unlike in (5.3a, b, c), here J^{μ} is the non-zero current given by

$$J^{\mu} = qnu^{\mu}, \tag{6.63}$$

where q is the charge, n the number and u^{μ} the four-velocity of matter. As in the Newton–Maxwell theory the mass m and charge q of the particles are constant while n is a function of position. As in (4.1) $T^{\mu\nu}$, the energy-momentum tensor of matter, is given by

$$T^{\mu\nu} = mnu^{\mu}u^{\nu}. \tag{6.64}$$

Note that $E^{\mu}_{\mu} = 0$ and $T = T^{\mu}_{\mu} = mn$.

We verify the conservation of the total energy-momentum given by

$$(T^{\mu\nu} + E^{\mu\nu})_{;\nu} = 0. \tag{6.65}$$

With the use of (6.61b, c) and (6.62) one can show, through a standard procedure, that

$$E^{\mu\nu}{}_{;\nu} = -F^{\mu\nu}J_{\nu}. \tag{6.66}$$

With the use of the same procedure as we followed in (4.3) and (4.4), one can show that

$$T^{\mu\nu}{}_{;\nu} = mn\left(\frac{du^{\mu}}{ds} + \Gamma^{\mu}_{\nu\sigma}u^{\nu}u^{\sigma}\right) = mn\frac{Du^{\mu}}{ds}. \tag{6.67}$$

With the use of the following equation of motion for a charged particle in the field $F^{\mu\nu}$ (s is the proper time along the world-line of the particle)

$$m\frac{Du^{\mu}}{ds} = qF^{\mu\nu}u_{\nu}, \qquad (6.68)$$

and the definition (6.63) of J^{μ}, it is readily verified that (6.66), (6.67) and (6.68) imply that (6.65) is satisfied.

As before, we specialize to the general axially symmetric rotating metric

$$ds^2 = f\,dt^2 - 2k\,d\phi\,dt - l\,d\phi^2 - e^{\mu}(d\rho^2 + dz^2) \qquad (6.69)$$

where f, k, l and μ are functions of ρ and z only. Setting $(x^0, x^1, x^2, x^3) = (t, \rho, z, \phi)$, as in (4.6) we get

$$u^0 = \frac{dt}{ds} = (f - 2\Omega k - \Omega^2 l)^{-1/2}, \quad u^1 = u^2 = 0, \quad u^3 = \frac{d\phi}{ds} = \Omega u^0, \quad (6.70)$$

where Ω is the angular velocity, in general a function of ρ and z. $F_{\mu\nu}$ is defined by

$$F_{\mu\nu} = A_{\mu,\nu} - A_{\nu,\mu}, \qquad (6.71)$$

in terms of the four-vector potential A_{μ} (a comma denotes partial differentiation) as a consequence of which (6.61b) is satisfied identically. In the present coordinate system we take A_{μ} to be of the form

$$(A_0, A_1, A_2, A_3) = (\Phi, 0, 0, \chi) \qquad (6.72)$$

(Note that this is different to the definition (5.5)), where Φ and χ are related to the electric and magnetic potentials respectively and are functions of ρ and z only. The equation (6.61a) can be written as follows, with the use of (6.62) and (6.64):

$$R_{\mu\nu} = 8\pi m n(u_{\mu}u_{\nu} - \tfrac{1}{2}g_{\mu\nu}) - 2F_{\mu}{}^{\alpha}F_{\nu\alpha} + \tfrac{1}{2}g_{\mu\nu}F_{\alpha\beta}F^{\alpha\beta}. \qquad (6.73)$$

Since all functions are independent of t and ϕ the equation of conservation of matter $(nu^{\mu})_{;\mu} = 0$ is satisfied identically as are the t- and ϕ-components of (6.68). The other two components of (6.68) reduce to the following equations:

$$\tfrac{1}{2}m(f - 2\Omega k - \Omega^2 l)^{-1/2}(f_{\rho} - 2\Omega k_{\rho} - \Omega^2 l_{\rho}) = q(\Phi_{\rho} + \Omega\chi_{\rho}), \quad (6.74a)$$
$$\tfrac{1}{2}m(f - 2\Omega k - \Omega^2 l)^{-1/2}(f_z - 2\Omega k_z - \Omega^2 l_z) = q(\Phi_z + \Omega\chi_z). \quad (6.74b)$$

Three of the field equations (6.73) are as follows:

$$\begin{aligned}
2e^{\mu}D^{-1}R_{00} &= (D^{-1}f_{\rho})_{\rho} + (D^{-1}f_z)_z + D^{-3}f\Sigma \\
&= 8\pi m n D^{-1}e^{\mu}(f - 2\Omega k - \Omega^2 l)^{-1} \\
&\quad \times [2\Omega k(-f + \Omega k) + f(f + \Omega^2 l)] \\
&\quad + 4D^{-3}[(\tfrac{1}{2}fl + k^2)(\Phi_{\rho}^2 + \Phi_z^2) + kf(\Phi_{\rho}\chi_{\rho} + \Phi_z\chi_z) \\
&\quad + \tfrac{1}{2}f^2(\chi_{\rho}^2 + \chi_z^2)],
\end{aligned} \qquad (6.75a)$$

$$-2e^{\mu}D^{-1}R_{03} = (D^{-1}k_{\rho})_{\rho} + (D^{-1}k_z)_z + D^{-3}k\Sigma$$
$$= 8\pi mn D^{-1}e^{\mu}(f - 2\Omega k - \Omega^2 l)^{-1}(fk + 2\Omega fl - \Omega^2 kl)$$
$$- D^{-3}[\tfrac{1}{2}kl(\Phi_{\rho}^2 + \Phi_z^2) + fl(\Phi_{\rho}\chi_{\rho} + \Phi_z\chi_z)$$
$$- \tfrac{1}{2}kf(\chi_{\rho}^2 + \chi_z^2)], \qquad (6.75b)$$

$$-2e^{\mu}D^{-1}R_{33} = (D^{-1}l_{\rho})_{\rho} + (D^{-1}l_z)_z + D^{-3}l\Sigma$$
$$= -8\pi mn D^{-1}e^{\mu}(f - 2\Omega k - \Omega^2 l)^{-1}$$
$$\times (fl + 2k^2 + 2\Omega kl + \Omega^2 l^2)$$
$$- 4D^{-3}[\tfrac{1}{2}l^2(\Phi_{\rho}^2 + \Phi_z^2) - kl(\Phi_{\rho}\chi_{\rho} + \Phi_z\chi_z)$$
$$+ (\tfrac{1}{2}fl + k^2)(\chi_{\rho}^2 + \chi_z^2)], \qquad (6.75c)$$

where $D^2 = fl + k^2$ and $\Sigma \equiv f_{\rho}l_{\rho} + f_z l_z + k_{\rho}^2 + k_z^2$. An important combination of these equations is the following:

$$e^{\mu}D^{-1}(lR_{00} - 2kR_{03} - fR_{33}) = D_{\rho\rho} + D_{zz} = 0. \qquad (6.76)$$

Thus we get the same equation for D as in (2.3) although the corresponding equations (6.75a, b, c) are much more complicated than (2.2a, b, c). It follows that we can carry out the same procedure as that after (2.3) and choose a coordinate system such that f, l and k are related as follows (see (2.9)):

$$fl + k^2 = \rho^2. \qquad (6.77)$$

It is clear that the same procedure could have been carried out for the Einstein–Maxwell exterior equations given by (5.3a) since these can be derived from (6.75a, b, c) by setting $n = 0$. It is for this reason that we started in Section 5.2 with the metric (2.16) (see the equation after (5.4)), which assumes the relation (6.77) between f, k and l. Note that (6.77) is also valid for differential rotation with $\Omega = \Omega(\rho, z)$.

6.5. Solution for rigid rotation and vanishing Lorentz force

In this section we confine ourselves to a constant value of Ω, that is, to rigid rotation. As in Section 4.2 we transform to a system of coordinates rotating with angular velocity Ω to that given by (6.69) so that the functions l, k, f, Φ transform to L, K, F, Ψ given by

$$L = l, \quad K = k + \Omega l, \quad F = f - 2\Omega k - \Omega^2 l, \quad \Psi = \Phi + \Omega \chi. \qquad (6.78)$$

The relation (6.77) is invariant as before, that is,

$$FL + K^2 = \rho^2. \qquad (6.79)$$

Note that because of (6.77) or (6.79) only two of the field equations (6.75a, b, c) are independent. The rest of the independent field equations can be written as follows (with $D = \rho$):

$$-2e^{\mu}R_{33} = \Delta L + \rho^{-2}L(F_{\rho}L_{\rho} + F_z L_z + K_{\rho}^2 + K_z^2)$$

$$= -8\pi mne^{\mu}F^{-1}(FL + 2K^2)$$

$$-4\rho^{-2}[\tfrac{1}{2}L^2(\Psi_{\rho}^2 + \Psi_z^2) - LK(\Psi_{\rho}\chi_{\rho} + \Psi_z\chi_z)$$

$$+(\tfrac{1}{2}FL + K^2)(\chi_{\rho}^2 + \chi_z^2)], \qquad (6.80a)$$

$$2e^{\mu}\rho[(k + \Omega l)R_0^3 + \Omega(f - \Omega k)R_3^0](K - \Omega L)^{-1}$$

$$= [\rho^{-1}(FK_{\rho} - KF_{\rho})]_{\rho} + [\rho^{-1}(FK_z - KZ_z)]_z$$

$$= -4\rho^{-1}K(\Psi_{\rho}^2 + \Psi_z^2) - 4\rho^{-1}F(\Psi_{\rho}\chi_{\rho} + \Psi_z\chi_z), \qquad (6.80b)$$

$$-e^{\mu}R_1^1 = -\tfrac{1}{2}(\mu_{\rho\rho} + \mu_{zz}) + \tfrac{1}{2}\rho^{-1}\mu_{\rho} + \tfrac{1}{2}\rho^{-2}(F_{\rho}L_{\rho} + K_{\rho}^2)$$

$$= 4\pi mne^{\mu} + \rho^{-2}[L(\Psi_z^2 - \Psi_{\rho}^2) + 2K(\Psi_{\rho}\chi_{\rho} - \Psi_z\chi_z)$$

$$+ F(\chi_{\rho}^2 - \chi_z^2)], \qquad (6.80c)$$

$$-e^{\mu}R_2^2 = -\tfrac{1}{2}(\mu_{\rho\rho} + \mu_{zz}) - \tfrac{1}{2}\rho^{-1}\mu_{\rho} + \tfrac{1}{2}\rho^{-2}(F_z L_z + K_z^2),$$

$$= 4\pi mne^{\mu} + \rho^{-2}[L(\Psi_{\rho}^2 - \Psi_z^2) - 2K(\Psi_{\rho}\chi_{\rho} - \Psi_z\chi_z)$$

$$+ F(\chi_z^2 - \chi_{\rho}^2)], \qquad (6.80d)$$

$$R_{12} = \tfrac{1}{2}\rho^{-1}\mu_z + \tfrac{1}{4}\rho^{-2}(F_{\rho}L_z + F_z L_{\rho} + 2K_{\rho}K_z)$$

$$= 2\rho^{-2}[-L\Psi_{\rho}\Psi_z + K(\Psi_{\rho}\chi_z + \Psi_z\chi_{\rho}) + F\chi_{\rho}\chi_z]. \qquad (6.80e)$$

Note that in the new functions F, L, K, Ψ the equations (6.80*a–e*) are independent of Ω, which is reasonable since these equations are in terms of the metric of the rotated coordinate system in which the matter is at rest. The ρ- and z-components of the Maxwell equation (6.61*c*) vanish identically, while the t- and ϕ-components are as follows:

$$(-L\Psi_{\rho} + K\chi_{\rho})_{\rho} + (-L\Psi_z + K\chi_z)_z - \rho^{-1}(-L\Psi_{\rho} + K\chi_{\rho})$$

$$= -4\pi\rho^2 e^{\mu}J^0 = -4\pi q\rho^2 e^{\mu}nF^{-1/2}, \qquad (6.81a)$$

$$(K\Psi_{\rho} + F\chi_{\rho})_{\rho} + (K\Psi_z + F\chi_z)_z - \rho^{-1}(K\Psi_{\rho} + F\chi_{\rho})$$

$$= 4\pi\rho^2 e^{\mu}(J^3 - \Omega J^0) = 0. \qquad (6.81b)$$

Equations (6.80*a–e*) and (6.81*a, b*) are to be solved for the four unknowns Ψ, χ, μ, n and two of F, L, K. In addition to (6.80*a–e*) and (6.81*a, b*) one has (6.74*a, b*) which in the case of the constant Ω can be integrated to yield

$$f - 2\Omega k - \Omega^2 l = [qm^{-1}(\Phi + \Omega\chi) + a]^2, \qquad (6.82)$$

where a is an arbitrary constant. In terms of F and Ψ (6.82) can be written as

$$F = (q'\Psi + a)^2, \qquad (6.83)$$

where $q' = qm^{-1}$.

We now proceed to find a solution of the systems (6.80*a–e*) (6.81*a, b*) and (6.83). To begin with we make the assumption that F is constant like the Van Stockum interior solution (see (4.14)). From (6.83) it follows that Ψ is also constant. Thus we have

$$F = \text{constant} = F_0, \quad \Psi = \text{constant} = \Psi_0. \qquad (6.84)$$

It follows from (6.80b) that K satisfies the same equation as in the Van Stockum solution (see (4.15)):

$$\Delta K = K_{\rho\rho} + K_{zz} - \rho^{-1}K_{\rho} = 0, \tag{6.85}$$

which, as before, can be solved as follows:

$$K = \alpha\xi, \quad \xi = \rho\eta_{\rho}, \quad \Delta\xi = 0, \tag{6.86}$$

where η is harmonic, and α is an arbitrary constant which, if necessary, can be absorbed into the definition of ξ. Equation (6.80a) then reduces to the following equation:

$$\Delta L + \alpha^2\rho^{-2}L(\xi_{\rho}^2 + \xi_z^2) = -8\pi mne^{\mu}(L + 2F_0^{-1}\alpha^2\xi^2)$$
$$- 2\rho^{-2}(L + 2F_0^{-1}\alpha^2\xi^2)F_0(\chi_{\rho}^2 + \chi_z^2). \tag{6.87}$$

(Compare this with (4.18).) As in (4.19) and (4.20) we get

$$L = F^{-1}(\rho^2 - K^2) = F_0^{-1}(\rho^2 - \alpha^2\xi^2) \tag{6.88}$$

and

$$\Delta L = -2F_0^{-1}\alpha^2(\xi_{\rho}^2 + \xi_z^2). \tag{6.89}$$

From (6.81b), with the use of (6.84), we see at once that

$$\Delta\chi = \chi_{\rho\rho} + \chi_{zz} - \rho^{-1}\chi_{\rho} = 0, \tag{6.90}$$

so that we can take

$$\chi = \beta\xi, \tag{6.91}$$

where ξ is given by (6.86) and β is another arbitrary constant, which, however, cannot be set equal to unity if we decide to take $\alpha = 1$. Substituting from (6.89) and (6.91) into (6.87) we find a factor $(1 + \alpha^2\rho^{-2}\xi^2)$ throughout, after cancellation of which we get the following equation:

$$8\pi mn = (\alpha^2 - 2F_0\beta^2)\rho^{-2}e^{-\mu}(\xi_{\rho}^2 + \xi_z^2). \tag{6.92}$$

This equation reduces to (4.21) if $\beta = 0$. From (6.80c, d, e) we find that μ is given by the following two equations:

$$\mu_{\rho} = \tfrac{1}{2}\rho^{-1}(\alpha^2 - 4F_0\beta^2)(\xi_z^2 - \xi_{\rho}^2), \quad \mu_z = \rho^{-1}(-\alpha^2 + 4F_0\beta^2)\xi_{\rho}\xi_z. \tag{6.93}$$

the consistency of which is guaranteed by $\Delta\xi = 0$ (see (4.24)). Equation (6.81a) now reduces to

$$4\pi qn = -\alpha\beta F_0^{1/2}\rho^{-2}e^{-\mu}(\xi_{\rho}^2 + \xi_z^2). \tag{6.94}$$

Comparing with (6.92) we find that q, m, α, β and F_0 must satisfy

$$2\alpha\beta F_0^{1/2}m + q(\alpha^2 - 2F_0\beta^2) = 0. \tag{6.95}$$

From (6.92) and (6.95) we see that, assuming $e^{-\mu}$ to be positive, we can always arrange the number density n given by (6.92) to be positive by choosing the product $\alpha\beta q$ to be negative. In particular, taking α to be always positive, if q is positive the coefficient β of the magnetic potential χ given by (6.91) is negative, whereas the opposite is true if q is negative. Our solution reduces to the Van Stockum solution if $\beta = 0$ (i.e. if $q = 0$), that is, when the dust is uncharged so that there is no electromagnetic field.

Examination of the equations of motion (6.74a, b) reveals that the left hand sides of the equations represent the gravitational and centrifugal forces on the dust particles whereas the right hand sides represent the electromagnetic or Lorentz forces. When (6.84) is satisfied both sides of these equations vanish separately. Thus the above solution represents a situation in which the Lorentz force on a typical dust particle vanishes. Comparing with the Newtonian situation given by (6.60a, b) we see that in the latter the gravitational potential V is necessarily independent of z, whereas this is not true in general relativity. The gravitational potential corresponds to the metric in general relativity and the metric for the above solution has a non-trivial z-dependence through the harmonic function η (see (6.86)). Thus the general relativistic situation gives more varied possible configurations.

Looking at (6.74a, b), (6.78) and (6.84) we see that in the frame of reference in which the charged dust particles are at rest, the electric field vanishes, but there is a non-zero magnetic field which, however, exerts no force on the particles since the latter are at rest in that frame.

The solution given by (6.84), (6.86), (6.90), (6.91), (6.92), (6.93) and (6.95) is one of a class of solutions of the system (6.80$a-e$), (6.81a, b) and (6.83), namely, a class of solutions in which the Lorentz force vanishes. The general solution of this system of equations has not yet been found. This general solution would represent situations with non-vanishing Lorentz forces. Bonnor (1980b) has generalized the above solution to the case in which the charge to mass ratio of the dust particles is a function of position.

As in the case of neutral dust (see (4.25) and (4.26)) we can consider the cylindrically symmetric form of the above solution by choosing ξ as in (4.25). As before we can take $F_0 = 1$ without loss of generality, and the solution is given by

$$\left.\begin{aligned}
F &= 1, \quad K = \alpha\rho^2, \quad L = \rho^2(1 - \alpha^2\rho^2), \quad e^\mu = e^{-(\alpha^2 - 4\beta^2)\rho^2}, \\
n &= (2\pi)^{-1}(\alpha^2 - 2\beta^2)e^{(\alpha^2 - 4\beta^2)\rho^2}, \quad \chi = \beta\rho^2
\end{aligned}\right\} . \quad (6.96)$$

Since the Lorentz force vanishes, the dust particles follow geodesics and we can analyze the situation in a similar manner to that in Section 4.4.

6.6. Cylindrically symmetric global solution

Equation (6.96) gives cylindrically symmetric solution for rigidly rotating
charged dust for which the Lorentz force vanishes. We now look for
cylindrically symmetric solutions for rigid rotation (that is, for constant Ω)
for which the Lorentz force does not necessarily vanish. We shall find such
an interior solution and also find an exterior (electrovac) solution which
will match smoothly onto the interior solution and thus get a global
solution; as mentioned earlier, this is rare either for the Einstein or the
Einstein–Maxwell equations.

The cylindrically symmetric interior Einstein–Maxwell equations are
given as follows:

$$F_{\rho\rho} - \rho^{-1}F_\rho + \rho^{-2}F(F_\rho L_\rho + K_\rho^2)$$
$$= 8\pi mne^\mu F + 4\rho^{-2}[(\tfrac{1}{2}LF + K^2)\Psi_\rho^2 + KF\Psi_\rho\chi_\rho + \tfrac{1}{2}F^2\chi_\rho^2], \quad (6.97a)$$
$$K_{\rho\rho} - \rho^{-1}K_\rho + \rho^{-2}K(F_\rho L_\rho + K_\rho^2)$$
$$= 8\pi mne^\mu K - 4\rho^{-2}[\tfrac{1}{2}LK\Psi_\rho^2 + LF\Psi_\rho\chi_\rho - \tfrac{1}{2}FK\chi_\rho^2], \quad (6.97b)$$
$$-\tfrac{1}{2}\mu_{\rho\rho} + \tfrac{1}{2}\rho^{-1}\mu_\rho + \tfrac{1}{2}\rho^{-2}(F_\rho L_\rho + K_\rho^2)$$
$$= 8\pi mne^\mu + \rho^{-2}(-L\Psi_\rho^2 + 2K\Psi_\rho\chi_\rho + F\chi_\rho^2), \quad (6.97c)$$
$$-\tfrac{1}{2}\mu_{\rho\rho} - \tfrac{1}{2}\rho^{-1}\mu_\rho = 8\pi mne^\mu - \rho^{-2}(-L\Psi_\rho^2 + 2K\Psi_\rho\chi_\rho + F\chi_\rho^2), \quad (6.97d)$$
$$-L(\Psi_{\rho\rho} - \rho^{-1}\Psi_\rho) + K(\chi_{\rho\rho} - \rho^{-1}\chi_\rho) - L_\rho\Psi_\rho + K_\rho\chi_\rho = -4\pi q\rho^2 ne^\mu F^{-1/2},$$
$$\quad (6.97e)$$
$$K(\Psi_{\rho\rho} - \rho^{-1}\Psi_\rho) + F(\chi_{\rho\rho} - \rho^{-1}\chi_\rho) + K_\rho\Psi_\rho + F_\rho\chi_\rho = 0 \quad (6.97f)$$
$$F = (q'\Psi + a)^2. \quad (6.97g)$$

In Islam (1978a) and following papers we took the arbitrary constant a in
(6.97g) to be zero. This does not make any difference unless we want to take
the limit $q = 0$. Equations (6.97a, b) are certain combinations of (6.80a, b)
and equations of (6.97c, d) are derived from (6.80c, d) by making all
functions independent of z. Similarly (6.97e, f) are obtained from (6.81a, b),
while (6.97g) is just (6.83) written here again for convenience.

After division by ρ (6.96f) can be integrated to give

$$K\Psi_\rho + F\chi_\rho = A_\rho \quad (6.98)$$

where A is an arbitrary constant. Eliminating ne^μ from (6.97a, b) we get an
equation which can be integrated as follows:

$$KF_\rho - FK_\rho = \rho(4A\Psi + A_0), \quad (6.99)$$

where A_0 is another arbitrary constant. Recall that F, K, L satisfy

$$FL + K^2 = \rho^2. \quad (6.100)$$

With the use of (6.97g), (6.98) and (6.99), equation (6.97e) can be reduced to the following equation:

$$-\tfrac{1}{2}mF_{\rho\rho} - \tfrac{1}{2}m\rho^{-1}F_\rho + \tfrac{3}{4}mF^{-1}F_\rho^2 - AA_0'qF^{-1/2} - 4mA^2 = -4\pi q^2 ne^\mu F,$$
(6.101)

where $A_0' = A_0 - 4Aq'^{-1}a$. With the use of (6.97a), (6.98), (6.99), (6.101) and (6.110) below one can obtain the following single differential equation for F:

$$(q^2 - m^2)(F_{\rho\rho} + \rho^{-1}F_\rho - F^{-1}F_\rho^2) + 2(4m^2 - q^2)A^2$$
$$+ 6qmAA_0'F^{-1/2} + A_0'^2q^2F^{-1} = 0.$$
(6.102)

Once (6.102) has been solved for F, the other functions K, Ψ, χ, μ and n can be obtained with the use of (6.97c, d), (6.98) and (6.99).

We set $A = 0$ in (6.102). Then it can be seen that the following is a solution of the resulting equation:

$$F = \xi_0 + \eta_0\rho^2,$$
(6.103)

where ξ_0 and η_0 are constants related as follows:

$$4\xi_0\eta_0(q^2 - m^2) + q^2A_0'^2 = 0.$$
(6.104)

We now determine the functions K, Ψ, χ, μ and n corresponding to the solution (6.103).

From (6.97g) we get

$$q'\Psi + a = \pm(\xi_0 + \eta_0\rho^2)^{1/2},$$
(6.105)

where we choose the positive sign for convenience, as we did in deriving (6.98), (6.99) and (6.102). Equation (6.99) (with $A = 0$) can be solved for K to yield (we set $a = 0$ for convenience)

$$K = (2\eta_0)^{-1}A_0 + C(\xi_0 + \eta_0\rho^2),$$
(6.106)

where C is an arbitrary constant. Equation (6.98) with $A = 0$ can then be solved to yield

$$\chi = (2\eta_0q)^{-1}mA_0(\xi_0 + \eta_0\rho^2)^{-1/2} - q^{-1}mC(\xi_0 + \eta_0\rho^2)^{1/2}, \quad (6.107)$$

where we have set a trivial additive arbitrary constant equal to zero. From (6.101) (with $A = 0$) we get

$$4\pi ne^\mu = m\eta_0q^{-2}(2\xi_0 - \eta_0\rho^2)(\xi_0 + \eta_0\rho^2)^{-2}.$$
(6.108)

Next we subtract (6.97d) from (6.97c) to obtain

$$\mu_\rho = -\tfrac{1}{2}\rho^{-1}(F_\rho L_\rho + K_\rho^2) + 2\rho^{-1}(-L\Psi_\rho^2 + 2K\Psi_\rho\chi_\rho + F\chi_\rho^2).$$
(6.109)

With the use of (6.100) we derive the identity

$$F_\rho L_\rho + K_\rho^2 = -\rho^{-2}F^{-2}F_\rho^2 + 2\rho F^{-1}F_\rho + F^{-2}(KF_\rho - FK_\rho)^2$$
$$= -\rho^{-2}F^{-2}F_\rho^2 + 2\rho F^{-1}F_\rho + \rho^2 F^{-2}(4A\Psi + A_0)^2, \quad (6.110)$$

where in the last step we have used (6.99). With the use of (6.110) and expressions for Ψ, χ, F and K, (6.109) yields the following integral:

$$e^\mu = \alpha(\xi_0 + \eta_0\rho^2)^{-m^2q-2}, \qquad (6.111)$$

where α is an arbitrary constant. Equations (6.108) and (6.111) lead to the following expression for the number density n:

$$n = (4\pi q^2 \alpha)^{-1} m \eta_0 (2\xi_0 - \eta_0\rho^2)(\xi_0 + \eta_0\rho^2)^{-2+m^2q-2}. \qquad (6.112)$$

The two equations (6.97c, d) lead to (6.109). It is thus adequate to verify that one of (6.97c, d) is satisfied by the above solution. It can be verified that (6.97d) is indeed satisfied. Thus (6.103)–(6.108), (6.111) and (6.112) constitute a solution of the system of equations given by (6.97a–g). If the solution is to be regular on the axis $\rho = 0$, the function L must be of order ρ^2 there, so that the geometry in the two-spaces $t =$ constant, $z =$ constant be Euclidean. (Recall that at any regular point of space–time it is possible to introduce a coordinate system so that the space–time is locally Minkowskian at the point.) Since $L = F^{-1}(\rho^2 - K^2)$, it follows that the constant term in K must vanish. This implies that

$$A_0 = -2C\xi_0\eta_0. \qquad (6.113)$$

This leads to a reduction in the number of constants in the solution.

Consider values of ξ_0, η_0 with $\xi_0 > 0$, $\eta > 0$. Then (6.104) implies that $q^2 < m^2$. For these values of ξ_0, η_0 the solution is well behaved everywhere inside the region where n (the number density) is non-zero and positive, that is, inside the region $\rho^2 < 2\xi_0/\eta_0$. The dust distribution has a boundary at $\rho^2 = 2\xi_0/\eta_0$, where the number density vanishes. The vacuum solution has to match smoothly for some value of ρ such that $\rho^2 < 2\xi_0/\eta_0$. From (6.111) and (6.112) it is clear that the presence of charge is essential for the solution, that is, the solution does not have a suitable limit as q tends to zero, since e^μ tends to zero and n tends to infinity. We had to choose $q^2 < m^2$ to obtain a regular solution. Recall that we encountered a similar situation in the cylindrically symmetrical solution (6.21a–e) in Newton–Maxwell theory. This is also true of another solution mentioned at the end of this section. Presumably in general relativity, as in Newtonian physics, equilibrium is not possible for $q^2 > m^2$ as the electromagnetic forces overwhelm the gravitational and centrifugal forces.

We now obtain an exterior (electrovac) solution which we will match smoothly on to the above solution. The exterior equations are derived by putting $n = 0$ in (6.97a–e) and ignoring (6.97g) which is not valid outside matter. We write these here for convenience, but remember that these equations are in terms of L, K, F so they refer to a rotating coordinate system:

$$F_{\rho\rho} - \rho^{-1}F_\rho + \rho^{-2}(F_\rho L_\rho + K_\rho^2)F$$
$$= 4\rho^{-2}[(\tfrac{1}{2}LF + K^2)\Psi_\rho^2 + KF\Psi_\rho\chi_\rho + \tfrac{1}{2}F^2\chi_\rho^2], \quad (6.114a)$$

$$K_{\rho\rho} - \rho^{-1}K_\rho + \rho^{-2}(F_\rho L_\rho + K_\rho^2)K$$
$$= -4\rho^{-2}(\tfrac{1}{2}LK\Psi_\rho^2 + LF\Psi_\rho\chi_\rho - \tfrac{1}{2}FK\chi_\rho^2), \quad (6.114b)$$

$$-\tfrac{1}{2}\mu_{\rho\rho} + \tfrac{1}{2}\rho^{-1}\mu_\rho + \tfrac{1}{2}\rho^{-2}(F_\rho L_\rho + K_\rho^2) = \rho^{-2}(-L\Psi_\rho^2 + 2K\Psi_\rho\chi_\rho + F\chi_\rho^2),$$
$$(6.114c)$$

$$-\tfrac{1}{2}\mu_{\rho\rho} - \tfrac{1}{2}\rho^{-1}\mu_\rho = -\rho^{-2}(-L\Psi_\rho^2 + 2K\Psi_\rho\chi_\rho + F\chi_\rho^2), \quad (6.114d)$$

$$-L(\Psi_{\rho\rho} - \rho^{-1}\Psi_\rho) + K(\chi_{\rho\rho} - \rho^{-1}\chi_\rho) - L_\rho\Psi_\rho + K_\rho\chi_\rho = 0, \quad (6.114e)$$

$$K(\Psi_{\rho\rho} - \rho^{-1}\Psi_\rho) + F(\chi_{\rho\rho} - \rho^{-1}\chi_\rho) + K_\rho\Psi_\rho + F_\rho\chi_\rho = 0. \quad (6.114f)$$

Equation (6.114f) can be integrated to give

$$K\Psi_\rho + F\chi_\rho = a'\rho \quad (6.115)$$

where a' is an arbitrary constant. Subtracting F times (6.114b) from K times (6.114a) we get an equation which integrates to

$$KF_\rho - FK_\rho = \rho(4a'\Psi + a_0), \quad (6.116)$$

where a_0 is another arbitrary constant. Equation (6.114e) can be integrated to give

$$-L\Psi_\rho + K\chi_\rho = a_1\rho, \quad (6.117)$$

where a_1 is another arbitrary constant. From (6.100), (6.115) and (6.117) we get

$$\Psi_\rho = \rho^{-1}(a'K - a_1 F). \quad (6.118)$$

With the use of (6.118) and relations similar to (6.110), equation (6.114a) can be written as

$$\rho FF_{\rho\rho} + FF_\rho - \rho F_\rho^2 - 2\rho F(\Psi_\rho^2 + a'^2) + \rho(4a'\Psi + a_0)^2 = 0. \quad (6.119)$$

With the use of (6.100), (6.115), (6.118) and (6.110), equation (6.114b) becomes, after some reduction,

$$F^2(\rho^2\Psi_{\rho\rho\rho} + \rho\Psi_{\rho\rho} - \Psi_\rho + a_1\rho F_{\rho\rho} - a_1 F_\rho)$$
$$+ (\rho\Psi_\rho + a_1 F)[-\rho F_\rho^2 + 2FF_\rho + \rho(4a'\Psi + a_0)^2] \quad (6.120)$$
$$+ 2\rho F(a'^2\rho\Psi_\rho - \rho\Psi_\rho^3 - a_1 F\Psi_\rho^2 - a'^2 a_1 F) = 0.$$

With the use of (6.119) this can be reduced to

$$F(\rho\Psi_{\rho\rho\rho} + \Psi_{\rho\rho} - \rho^{-1}\Psi_\rho) + \Psi_\rho(4a'^2\rho + F_\rho - \rho F_{\rho\rho}) = 0. \quad (6.121)$$

Equations (6.119) and (6.121) are the basic equations. Once these are solved the other functions can be derived from this solution.

Rotating charged dust

We now proceed to find a solution with $a' = 0$. From (6.118) we obtain

$$\rho \Psi_\rho = - a_1 F. \tag{6.122}$$

It can be verified that (6.121) with $a' = 0$ is satisfied identically by F and Ψ which are connected by (6.122). Equation (6.119) reduces to

$$\rho F F_{\rho\rho} + F F_\rho - \rho F_\rho^2 - 2a_1^2 \rho^{-1} F^3 + \rho a_0^2 = 0. \tag{6.123}$$

A solution of (6.123) is given as follows:

$$F = \sigma_0 \rho^{2/3}, \quad \sigma_0 = (a_0^2/2a_1^2)^{1/3}. \tag{6.124}$$

From (6.115), (6.116) and (6.122) the corresponding Ψ, K, χ can be determined as follows:

$$\Psi = -\tfrac{3}{2} a_1 \sigma_0 \rho^{2/3} + \Psi_0, \tag{6.125a}$$

$$K = - (3a_0/2\sigma_0)\rho^{4/3} - K_0 \sigma_0 \rho^{2/3}, \tag{6.125b}$$

$$\chi = - (9a_0 a_1/8\sigma_0)\rho^{4/3} - \tfrac{3}{2} K_0 a_1 \sigma_0 \rho^{2/3} + \chi_0, \tag{6.125c}$$

where Ψ_0, K_0, χ_0 are arbitrary constants. Both (6.114c, d) are satisfied by the fullowing μ:

$$e^\mu = \lambda \rho^{-4/9} \exp(-\tfrac{9}{2} a_1^2 \sigma_0 \rho^{2/3}), \tag{6.126}$$

where λ is an arbitrary constant. Thus (6.124), (6.125a, b, c) and (6.126) constitute an exact solution of the system (6.114a–f). This new solution was found by the author (Islam 1983b) and was also discovered independently by Van den Bergh and Wils (1983) and by Jordan (1983). We now proceed to match this solution to the interior solution found earlier in this section.

The F and K for the interior solution given by (6.103) and (6.106) satisfy

$$K F_\rho - F K_\rho = \rho A_0. \tag{6.127}$$

Since F, K and their derivatives must be continuous across the boundary, say at $\rho = \rho_0$, it is clear from (6.116) and (6.127) that we must have

$$A_0 = a_0. \tag{6.128}$$

From (6.103) and (6.124) we see that for F and F_ρ to be continuous across $\rho = \rho_0$ the following relations hold:

$$\xi_0 = \tfrac{2}{3}\sigma_0 \rho_0^{2/3}, \quad \eta_0 = \tfrac{1}{3}\sigma_0 \rho_0^{-4/3}. \tag{6.129}$$

We find from (6.125b) and (6.106) with (6.113) that K and K_ρ are continuous if K_0 is given as follows:

$$K_0 = - (3a_0/4\sigma_0^2)\rho_0^{2/3}. \tag{6.130}$$

From (6.111) and (6.126) it can be shown that μ_ρ is continuous if the following relation is satisfied:

$$- (3a_0^2/2\sigma_0^2)\rho_0^{2/3} q^2 - \tfrac{4}{9} q^2 + \tfrac{2}{3} m^2 = 0. \tag{6.131}$$

The condition (6.104) can be written as

$$(a_0^2/\sigma_0^2)\rho_0^{2/3}q^2 + \tfrac{8}{9}(q^2 - m^2) = 0. \tag{6.132}$$

Equations (6.131) and (6.132) imply the following:

$$4q^2 = 3m^2, \quad 27a_0^2\rho_0^{2/3} = 8\sigma_0^2. \tag{6.133}$$

The continuity of μ relates and α and λ as follows:

$$\alpha = e^{-2/3}\sigma_0^{4/3}\rho_0^{4/9}\lambda. \tag{6.134}$$

The functions Ψ and χ can be made continuous trivially by adjusting the constants Ψ_0 and χ_0 in (6.125a, c). From (6.105) (with the positive sign and (6.125)) it can be shown that Ψ_ρ is continuous if (6.133) is satisfied with $q = -\tfrac{1}{2}(3)^{1/2}m$. The minus sign occurs here because we chose the positive sign in (6.105). With the choice of the negative sign in (6.105) one gets a positive q. Finally, it can be verified from (6.107) and (6.125c) that χ_ρ is continuous if again (6.133) is satisfied, with $q = -\tfrac{1}{2}(3)^{1/2}m$. Thus all the functions F, K, μ, Ψ and χ and their derivatives are continuous across $\rho = \rho_0$ so that the interior solution matches smoothly onto the exterior one. The number density of particles n can be written as follows:

$$n = (3\pi m\alpha)^{-1}\eta_0^{-2/3}(4\rho_0^2 - \rho^2)(2\rho_0^2 + \rho^2)^{-2/3}, \tag{6.135}$$

where η_0 is given by (6.129). Thus n is positive and well behaved inside the matter distribution $\rho \leqslant \rho_0$.

Before considering some properties of the solution in the next section, we mention another interior solution (Islam 1983a). This solution is obtained by setting $A_0 = 0$ in (6.102). The solution is as follows:

$$F = a''^2[1 - (\beta'/8a''^2)\rho^2]^2 \tag{6.136a}$$

where a'' is an arbitrary constant and

$$\beta' = 2(4m^2 - q^2)A^2/(q^2 - m^2), \tag{6.136b}$$

$$\Psi = mq^{-1}(1 - \frac{\beta'}{8a''^2}\rho^2), \tag{6.136c}$$

$$K = -2Am(a''q)^{-1}(1 - \frac{\beta'}{16a''^2}\rho^2)\rho^2. \tag{6.136d}$$

Here an arbitrary constant has been adjusted so that K is proportional to ρ^2 as ρ tends to zero, to avoid a singularity on the axis of symmetry $\rho = 0$;

$$\chi = 4A\beta'^{-1}\left(1 - \frac{2m^2}{q^2}\right)\left(1 - \frac{\beta'}{8a''^2}\rho^2\right)^{-1} + \frac{Am^2}{q^2a''^2}\rho^2 + \chi_0' \tag{6.136e}$$

where χ_0' is an arbitrary constant. The function μ is given by

$$e^\mu = \alpha'[1 - (\beta'/8a''^2)\rho^2]^{2 - 4m^2q^{-2}}, \tag{6.136f}$$

where α' is another arbitrary constant. Finally, the number density is as follows:

$$n = (4\pi q^2 \alpha')^{-1} m \left[\frac{A^2(8m^2 - 5q^2)}{a''^2(m^2 - q^2)} - \frac{\beta'^2}{16a''^4} \rho^2 \right]$$

$$\times (1 - \frac{\beta'}{8a''^2}\rho^2)^{-4 + 4m^2 q^{-2}}. \qquad (6.136g)$$

Again the solution is well behaved, with the number density positive, inside a certain value of ρ for $q^2 < m^2$. We refer to Islam (1983a) for the derivation and properties of this solution. Other cylindrically symmetric interior solutions, which, however, have singularities on $\rho = 0$, can be found in Islam (1978a, 1984). We also refer to Van den Bergh and Wils (1984d) for generalization of some of the solutions given here.

6.7. Some properties of the global solution

We consider some properties of the global solution of the last section. The interior and exterior solutions are regular for all finite values of ρ with $\rho \leqslant \rho_0$ and $\rho > \rho_0$ respectively. However, e^μ tends to zero as ρ tends to infinity, so there is a possibility of a singularity at $\rho = \infty$. To examine this further, we compute a scalar curvature, one of the simplest ones being $F_{\mu\nu}$ $F^{\mu\nu}$, where $F_{\mu\nu}$ is given by (6.71). It can be shown that for the exterior solution above (see the end of this section)

$$F_{\mu\nu}F^{\mu\nu} = -(2a_1^2\sigma_0/\lambda)\rho^{-8/9}\exp(\tfrac{9}{2}a_1^2\sigma_0\rho^{2/3}) \qquad (6.137)$$

which tends to infinity as ρ tends to infinity, suggesting the presence of sources at $\rho = \infty$. However, we note the fact that the spatial distance from any finite value of ρ to $\rho = \infty$ along the curve $\phi = $ constant, $z = $ constant is finite (see Islam 1983b).

We now work out the expansion, shear and rotation of the interior part of the global solution. The covariant derivative of the covariant four-velocity u_μ can in general be decomposed as follows:

$$u_{\mu;\nu} = w_{\mu\nu} + \sigma_{\mu\nu} + \tfrac{1}{3}(g_{\mu\nu} - u_\mu u_\nu)u^\sigma_{;\sigma} + \dot{u}_\mu u_\nu, \qquad (6.138)$$

where $u^\sigma_{;\sigma}$ is the expansion, $\dot{u}_\mu$ is the covariant four-acceleration given by

$$\dot{u}_\mu = u_{\mu;\sigma}u^\rho, \qquad (6.139)$$

and $\omega_{\mu\nu}$ and $\sigma_{\mu\nu}$ are respectively the rotation and shear tensors given by

$$w_{\mu\nu} = \tfrac{1}{2}(u_{\mu;\nu} - u_{\nu;\mu}) - \tfrac{1}{2}(\dot{u}_\mu u_\nu - \dot{u}_\nu u_\mu), \qquad (6.140a)$$

$$\sigma_{\mu\nu} = \tfrac{1}{2}(u_{\mu;\nu} + u_{\nu;\mu}) - \tfrac{1}{2}(\dot{u}_\mu u_\nu + \dot{u}_\nu u_\mu) - \tfrac{1}{3}(g_{\mu\nu} - u_\mu u_\nu)u^\sigma_{;\sigma}. \qquad (6.140b)$$

(See, for example, Mismer, Thorne and Wheeler (1973), p. 566, where, however, there are some differences in sign since for them $u_\mu u^\mu = -1$ while for us $u_\mu u^\mu = 1$.) Since the interior solution is given in a coordinate system in which the rotating dust is at rest, we have

$$(u^0, u^1, u^2, u^3) = (F^{-1/2}, 0, 0, 0). \qquad (6.141)$$

The covariant components of the four-velocity are given by

$$(u_0, u_1, u_2, u_3) = (F^{1/2}, 0, 0, -F^{-1/2}K). \qquad (6.142)$$

With the use of the Christoffel symbols (1.98a–d) (with all z-derivatives set equal to zero) we see that the contravariant and covariant components of the acceleration are as follows:

$$(\dot u^0, \dot u^1, \dot u^2, \dot u^3) = (0, \tfrac{1}{2}e^{-\mu}F^{-1}F_\rho, 0, 0), \qquad (6.143a)$$

$$(\dot u_0, \dot u_1, \dot u_2, \dot u_3) = (0, -\tfrac{1}{2}F^{-1}F_\rho, 0, 0). \qquad (6.143b)$$

The non-zero members of the covariant derivative of the contravariant four-velocity are

$$u^0{}_{;1} = \tfrac{1}{2}\rho^{-2}F^{-3/2}K(FK_\rho - KF_\rho), \quad u^1{}_{;0} = \tfrac{1}{2}e^{-\mu}F_\rho F^{-1/2},$$

$$u^1{}_{;3} = -\tfrac{1}{2}e^{-\mu}F^{-1/2}K_\rho, \quad u^3{}_{;1} = \tfrac{1}{2}\rho^{-2}F^{-1/2}(FK_\rho - KF_\rho). \qquad (6.144)$$

The corresponding non-zero covariant components are

$$u_{1;0} = -\tfrac{1}{2}F^{-1/2}F_\rho, \quad u_{1;3} = \tfrac{1}{2}F^{-1/2}K_\rho,$$

$$u_{3;1} = -\tfrac{1}{2}F^{-3/2}(FK_\rho - KF_\rho). \qquad (6.145)$$

With the use of (6.140a, b) and the equations following, one can readily show that the expansion and shear vanish identically, as is to be expected for rigidly rotating dust, while the only non-zero components of the rotation tensor are

$$w_{13} = -w_{31} = \tfrac{1}{2}F^{-3/2}(FK_\rho - KF_\rho). \qquad (6.146)$$

An invariant measure of the rotation is given by the following quantity:

$$w^2 = w^{\mu\nu}w_{\mu\nu} = \rho^{-2}F^{-2}e^{-\mu}(FK_\rho - KF_\rho)^2 = a_0^2 F^{-2}e^{-\mu}, \qquad (6.147)$$

where in the last relation we have used (6.127) and (6.128). Thus the constant a_0 is related directly to the amount of rotation.

In the evaluation of (6.137), the non-zero covariant components of the electromagnetic field tensor $F_{\mu\nu}$ are as follows:

$$F_{01} = -F_{10} = \Psi_\rho, \quad F_{13} = -F_{31} = -\chi_\rho, \qquad (6.148)$$

while the contravariant components are given by

$$\left.\begin{array}{l} F^{01} = -F^{10} = \rho^{-2}e^{-\mu}(K\chi_\rho - L\Psi_\rho) = a_1\rho^{-1}e^{-\mu}, \\ F^{13} = -F^{31} = -\rho^{-2}e^{-\mu}(K\Psi_\rho + F\chi_\rho) = 0, \end{array}\right\} \qquad (6.149)$$

where we have used (6.117) and (6.115) (with $a' = 0$).

In Section 6.6 we found a global cylindrically symmetric solution for rotating charged dust in Newton–Maxwell theory, where all the functions were regular everywhere. In Section 6.6 we found a general relativistic global solution for values of the charge given by $q^2 = \frac{3}{4}m^2$. This solution, however, has a possible singularity at $\rho = \infty$. This illustrates the extreme difficulty of finding general relativistic analogues of even very simple physical situations in Newtonian physics. In spite of the possible singularity at $\rho = \infty$, we believe the global solution found in Section 6.6 is worth studying in detail. For example, one could get some insight into the solution by studying equations of motion of charged dust particles.

The problem of axially symmetric differential rotation in general relativity can be brought to a similar stage as in Newtonian–Maxwell theory considered in Section 6.3, namely, the problem can be reduced to three coupled equations which become determinate once an arbitrary function of the angular velocity $\Omega(\rho, z)$ is specified (see (6.37)). In general relativity we get a system of three equations instead of the single equation (6.37). This problem will be considered in the next section.

6.8. Field equations for differential rotation

As noted in Section 6.4, the three field equations (6.75a, b, c) and the relations (6.76) and (6.77) which follow from these are also valid for differential rotation, that is, in situations where Ω is a function of ρ and z, which we now assume to be the case. We again transform the functions l, k, f and Φ to L, K, F and Φ given by (6.78) but now, since Ω is a function of ρ and z, these latter functions cannot be regarded as those obtained by a rotation of the coordinate system. The functions F, L and K satisfy (6.79) as before. In terms of the new functions (6.74a, b) can be written as

$$\tfrac{1}{2}mF^{-1/2}(F_\rho + 2K\Omega_\rho) = q(\Psi_\rho - \Omega_\rho\chi),$$ (6.150a)
$$\tfrac{1}{2}mF^{-1/2}(F_z + 2K\Omega_z) = q(\Psi_z - \Omega_z\chi).$$ (6.150b)

Equations (6.75a, b, c) (with $D = \rho$) can be combined to give the following three equations:

$$\Delta F + 4(K_\rho\Omega_\rho + K_z\Omega_z) - 2L(\Omega_\rho^2 + \Omega_z^2) + 2K\Delta\Omega + \rho^{-2}F\Sigma'$$
$$= 8\pi mne^\mu F + 4\rho^{-2}[(\tfrac{1}{2}LF + K^2)(\Psi_\rho^2 + \Psi_z^2) + KF(\Psi_\rho\chi_\rho + \Psi_z\chi_z)$$
$$- KF\chi(\chi_\rho\Omega_\rho + \chi_z\Omega_z) - (FL + 2K^2)\chi(\Psi_\rho\Omega_\rho + \Psi_z\Omega_z)$$
$$+ (\tfrac{1}{2}LF + K^2)\chi^2(\Omega_\rho^2 + \Omega_z^2) + \tfrac{1}{2}F^2(\chi_\rho^2 + \chi_z^2)],$$ (6.151a)

$$\Delta K - L\Delta\Omega - 2(L_\rho\Omega_\rho + L_z\Omega_z) + \rho^{-2}K\Sigma'$$
$$= 8\pi mne^\mu K - 4\rho^{-2}[\tfrac{1}{2}LK(\Psi_\rho^2 + \Psi_z^2) - LK\chi(\Psi_\rho\Omega_\rho + \Psi_z\Omega_z)$$
$$+ \tfrac{1}{2}LK\chi^2(\Omega_\rho^2 + \Omega_z^2) + LF(\Psi_\rho\chi_\rho + \Psi_z\chi_z) - LF\chi(\chi_\rho\Omega_\rho + \chi_z\Omega_z)$$
$$- \tfrac{1}{2}FK(\chi_\rho^2 + \chi_z^2)],$$ (6.151b)

$$\Delta L + \rho^{-2}L\Sigma' = 8\pi mne^\mu F^{-1}(-FL - 2K^2) - 4\rho^{-2}[\tfrac{1}{2}L^2(\Psi_\rho^2 + \Psi_z^2)$$
$$- LK(\Psi_\rho\chi_\rho + \Psi_z\chi_z) - L^2\chi(\Psi_\rho\Omega_\rho + \Psi_z\Omega_z)$$
$$+ \tfrac{1}{2}L^2\chi^2(\Omega_\rho^2 + \Omega_z^2) + (\tfrac{1}{2}LF + K^2)(\chi_\rho^2 + \chi_z^2)$$
$$+ LK\chi(\chi_\rho\Omega_\rho + \chi_z\Omega_z)], \tag{6.151c}$$

where

$$\Sigma' \equiv F_\rho L_\rho + K_\rho^2 + F_z L_z + K_z^2 + 2\Omega_\rho(KL_\rho - LK_\rho)$$
$$+ 2\Omega_z(KL_z - LK_z) + L^2(\Omega_z^2 + \Omega_z^2). \tag{6.151d}$$

Because of (6.79) only two of (6.151a, b, c) are independent. The other three non-trivial components of (6.73) can be written as follows:

$$R_{11} = -\tfrac{1}{2}(\mu_{\rho\rho} + \mu_{zz}) + \tfrac{1}{2}\rho^{-1}\mu_\rho$$
$$+ \tfrac{1}{2}\rho^{-2}[F_\rho L_\rho + K_\rho^2 + 2\Omega_\rho(KL_\rho - LK_\rho) + L^2\Omega_\rho^2]$$
$$= 4\pi mne^\mu + \rho^{-1}\Lambda, \tag{6.152a}$$

$$R_{12} = \tfrac{1}{2}\rho^{-1}\mu_z + \tfrac{1}{4}\rho^{-2}[F_\rho L_z + F_z L_\rho + 2K_\rho K_z$$
$$+ 2\Omega_\rho(KL_z - LK_z) + 2\Omega_z(KL_\rho - LK_\rho) + 2L^2\Omega_\rho\Omega_z]$$
$$= 2\rho^{-2}[-L(\Psi_\rho - \chi\Omega_\rho)(\Psi_z - \chi\Omega_z)$$
$$+ K(\Psi_\rho\chi_z + \Psi_z\chi_\rho) - K\chi(\Omega_\rho\chi_z + \Omega_z\chi_\rho) + F\chi_\rho\chi_z], \tag{6.152b}$$

$$R_{22} = -\tfrac{1}{2}(\mu_{\rho\rho} + \mu_{zz}) - \tfrac{1}{2}\rho^{-1}\mu_\rho + \tfrac{1}{2}\rho^{-2}[F_z L_z + K_z^2 + L^2\Omega_z^2$$
$$+ 2\Omega_z(KL_z - LK_z)] = 4\pi mne^\mu - \rho^{-2}\Lambda, \tag{6.152c}$$

where

$$\Lambda \equiv L[-(\Psi_\rho - \chi\Omega_\rho)^2 + (\Psi_z - \chi\Omega_z)^2]$$
$$+ 2K[\chi_\rho(\Psi_\rho - \chi\Omega_\rho) - \chi_z(\Psi_z - \chi\Omega_z)] + F(\chi_\rho^2 - \chi_z^2). \tag{6.152d}$$

The Maxwell equation (6.61c) reduces to the following two equations:

$$- L\Delta\Psi + K\Delta\chi + L\chi\Delta\Omega - L_\rho\Psi_\rho - L_z\Psi_z + K_\rho\chi_\rho + K_z\chi_z$$
$$+ L(\chi_\rho\Omega_\rho + \chi_z\Omega_z) + \chi(L_\rho\Omega_\rho + L_z\Omega_z) = -4\pi q\rho^2 e^\mu nF^{-1/2}, \tag{6.153a}$$

$$K\Delta\Psi + F\Delta\chi - K\chi\Delta\Omega + K_\rho\Psi_\rho + K_z\Psi_z + F_\rho\chi_\rho + F_z\chi_z$$
$$- \chi(K_\rho\Omega_\rho + K_z\Omega_z) - L(\Omega_\rho\Psi_\rho + \Omega_z\Psi_z) + L\chi(\Omega_\rho^2 + \Omega_z^2) = 0. \tag{6.153b}$$

The function Ω occurs in (6.151a, b, c), (6.152a, b, c) and (6.153a, b) only through the derivatives Ω_ρ, Ω_z, $\Omega_{\rho\rho}$, Ω_{zz}. These can be eliminated with the use of (6.150a, b). Recall that only two of (6.151a, b, c) are independent. Assuming that the derivatives of Ω have been eliminated, the six unknown functions μ, n, Ψ, χ and two of F, K, L satisfy the seven coupled equations (6.152a, b, c), (6.153a, b) and two of (6.151a, b, c). Thus one of these equations must be redundant. From (6.152a, c) we get

$$\mu_\rho = \tfrac{1}{2}\rho^{-1}(F_z L_z + K_z^2 - F_\rho L_\rho - K_\rho^2) + \rho^{-1}\Omega_z(KL_z - LK_z)$$
$$- \rho^{-1}\Omega_\rho(KL_\rho - LK_\rho) + \tfrac{1}{2}\rho^{-1}L^2(\Omega_z^2 - \Omega_\rho^2) + 2\rho^{-1}\Lambda, \tag{6.154}$$

where Λ is given by (6.152d). In the Appendix of Islam (1978a) it is shown that any solution of the system (6.151a, b, c) (any two of these), (6.152b), (6.153a, b) and (6.154) necessarily satisfy (6.152a, c). Thus the two equations (6.152a, c) can be replaced by the single equation (6.154), so that we have the six functions μ, n, Ψ, χ and say F, K satisfying the six coupled equations (6.151a, b), (6.152b), (6.153a, b) and (6.154).

6.9. System of three equations for differential rotation

In this section we shall reduce the field equations for differential rotation to a system of three coupled equations for three unknown functions. This system of three equations contains an arbitrary function of one variable and becomes determinate once this function is specified. This function can be regarded as an arbitrary function of the angular velocity $\Omega(\rho, z)$. This situation is thus similar to that encountered in Newton–Maxwell theory (Section 6.3), where, however, we had the single equation (6.37).

From (6.152b) and (6.154) we eliminate μ to get the following equation:

$$
\begin{aligned}
&\tfrac{1}{2}L_z\,\Delta F + (\tfrac{1}{2}F_z + K\Omega_z)\Delta L + (K_z - L\Omega_z)\Delta K + 4(L\Psi_z - K\chi_z - L\chi\Omega_z)\Delta\Psi \\
&+ 4(-K\Psi_z - F\chi_z + K\chi\Omega_z)\Delta\chi + (KL_z - LK_z - 4L\chi\Psi_z + L^2\Omega_z \\
&+ 4K\chi\chi_z + 4L\chi^2\Omega_z)\Delta\Omega - 2L_z(\Psi_\rho^2 - \Psi_z^2) + 4L_\rho\Psi_\rho\Psi_z \\
&+ 4K_z(\Psi_\rho\chi_\rho - \Psi_z\chi_z) - 4K_\rho(\chi_\rho\Psi_z + \chi_z\Psi_\rho) + 2F_z(\chi_\rho^2 - \chi_z^2) \\
&- 4F_\rho\chi_\rho\chi_z + 2\Omega_\rho(K_\rho L_z - L_\rho K_z) + 4(L\chi_z + L_z\chi)(\Psi_\rho\Omega_\rho - \Psi_z\Omega_z) \\
&+ LL_z(\Omega_z^2 - \Omega_\rho^2) - 4(L\chi_\rho + \chi L_\rho)(\Psi_z\Omega_\rho + \Psi_\rho\Omega_z) + 2LL_\rho\Omega_\rho\Omega_z \\
&+ 4K\Omega_z(\chi_\rho^2 + \chi_z^2) + 4\chi K_z(\chi_z\Omega_z - \chi_\rho\Omega_\rho) + 4\chi K_\rho(\chi_z\Omega_\rho + \chi_\rho\Omega_z) \\
&+ 2\chi(\chi L_z + 2L\chi_z)(\Omega_z^2 - \Omega_\rho^2) + 4\chi(\chi L_\rho + 2L\chi_\rho)\Omega_\rho\Omega_z = 0. \qquad (6.155a)
\end{aligned}
$$

For convenience we write (6.153b) here again:

$$
\begin{aligned}
&K\,\Delta\Psi + F\,\Delta\chi - K\chi\,\Delta\Omega + K_\rho\Psi_\rho + K_z\Psi_z + F_\rho\chi_\rho + F_z\chi_z \\
&- \chi(K_\rho\Omega_\rho + K_z\Omega_z) - L(\Omega_\rho\Psi_\rho + \Omega_z\Psi_z) + L\chi(\Omega_\rho^2 + \Omega_z^2) = 0. \qquad (6.155b)
\end{aligned}
$$

Eliminating ne^μ from (6.151a, b) we get the following equation:

$$
\begin{aligned}
&K\,\Delta F - F\,\Delta K + (FL + 2K^2)\Delta\Omega + 4K(K_\rho\Omega_\rho + K_z\Omega_z) \\
&+ 2F(L_\rho\Omega_\rho + L_z\Omega_z) - 2LK(\Omega_\rho^2 + \Omega_z^2) - 4K(\Psi_\rho^2 + \Psi_z^2) \\
&- 4F(\Psi_\rho\chi_\rho + \Psi_z\chi_z) + 4F\chi(\Omega_\rho\chi_\rho + \Omega_z\chi_z) \\
&+ 8K\chi(\Psi_\rho\Omega_\rho + \Psi_z\Omega_z) - 4K\chi^2(\Omega_\rho^2 + \Omega_z^2) = 0. \qquad (6.155c)
\end{aligned}
$$

Eliminating ne^μ from (6.151a) and (6.153a) we get

$$
\begin{aligned}
&q[\tfrac{1}{4}\rho^2\,\Delta F + \rho^2(K_\rho\Omega_\rho + K_z\Omega_z) - \tfrac{1}{2}\rho^2L(\Omega_\rho^2 + \Omega_z^2) + \tfrac{1}{2}\rho^2K\,\Delta\Omega \\
&+ \tfrac{1}{4}F(F_\rho L_\rho + F_z L_z + K_\rho^2 + K_z^2) + \tfrac{1}{2}F\Omega_\rho(KL_z - LK_\rho) \\
&+ \tfrac{1}{2}F\Omega_z(KL_z - LK_z) + \tfrac{1}{4}FL^2(\Omega_\rho^2 + \Omega_z^2) - (\tfrac{1}{2}LF + K^2)(\Psi_\rho^2 + \Psi_z^2) \\
&- KF(\Psi_\rho\chi_\rho + \Psi_z\chi_z) + KF\chi(\chi_\rho\Omega_\rho + \chi_z\Omega_z)
\end{aligned}
$$

$$+ (FL + 2K^2)\chi(\Psi_\rho\Omega_\rho + \Psi_z\Omega_z) - (\tfrac{1}{2}LF + K^2)\chi^2(\Omega_\rho^2 + \Omega_z^2)$$
$$- \tfrac{1}{2}F^2(\chi_\rho^2 + \chi_z^2)] + \tfrac{1}{2}mF^{3/2}[-L\,\Delta\Psi + K\,\Delta\chi + L\chi\,\Delta\Omega - L_\rho\Psi_\rho$$
$$- L_z\Psi_z + K_\rho\chi_\rho + K_z\chi_z + L(\chi_\rho\Omega_\rho + \chi_z\Omega_z) + \chi(L_\rho\Omega_\rho + L_z\Omega_z)] = 0.$$

$$(6.155d)$$

The function Ω without the derivative does not appear in (6.155a–d). The first and second derivatives of Ω with respect to ρ and z that appear in (6.155a–d) can be eliminated with the use of (6.150a, b). Thus it would appear, remembering that L can be expressed in terms of F, K and ρ through (6.79), that (6.155a–d) is a system of four equations for the four unknowns F, K, Ψ and χ. However, (6.150a, b) contain information which has not been used in simply eliminating Ω_ρ and Ω_z from (6.155a–d). Once this information is used, the system (6.155a–d) can be reduced to three coupled equations for three unknowns, this system of three equations containing an arbitrary function of one variable, so that these equations become determinate once this arbitrary function is specified. This we now proceed to do.

Equations (6.150a, b) imply

$$\sigma_\rho\Omega_z = \sigma_z\Omega_\rho, \tag{6.156a}$$

where

$$\sigma \equiv mF^{1/2} - q\Psi. \tag{6.156b}$$

The solution to (6.156a) is the following:

$$\Omega = \Omega(\sigma), \tag{6.157}$$

that is, Ω is a function of σ. Equations (6.156b), (6.157) and either one of (6.150a, b) then imply that

$$F = m^{-2}(q\Psi + \sigma)^2, \tag{6.158a}$$
$$K = -m^{-2}(q\chi + \Omega'^{-1})(q\Psi + \sigma), \tag{6.158b}$$

where a prime on Ω denotes differentiation with respect to σ: $\Omega' \equiv d\Omega/d\sigma$. With the use of (6.79), (6.157) and (6.158a, b), equations (6.155a–d) can be written entirely in terms of the three unknown functions σ, Ψ and χ as follows:

$$a_1\Delta\sigma + b_1\Delta\Psi + c_1\Delta\chi + d_1 = 0, \tag{6.159a}$$
$$a_2\Delta\sigma + b_2\Delta\Psi + c_2\Delta\chi + d_2 = 0, \tag{6.159b}$$
$$a_3\Delta\sigma + b_3\Delta\Psi + c_3\Delta\chi + d_3 = 0, \tag{6.159c}$$
$$a_4\Delta\sigma + b_4\Delta\Psi + c_4\Delta\chi + d_4 = 0, \tag{6.159d}$$

where the a_i, b_i, c_i, d_i ($i = 1, 2, 3, 4$) are functions of ρ, σ, Ψ, χ and the first derivatives of σ, Ψ and χ with respect to ρ and z. These functions can be read off from (6.155a–d) and are given explicitly in Appendix B of Islam (1979). Equations (6.159a–d) are four equations for the three unknowns σ, Ψ, χ. These equations can also be considered as algebraic equations for the three unknowns $\Delta\sigma$, $\Delta\Psi$, $\Delta\chi$ in terms of a_i, b_i, c_i, and d_i. Consistency demands that the following be true:

$$
\begin{vmatrix}
a_1 & b_1 & c_1 & d_1 \\
a_2 & b_2 & c_2 & d_2 \\
a_3 & b_3 & c_3 & d_3 \\
a_4 & b_4 & c_4 & d_4
\end{vmatrix} = 0. \tag{6.160}
$$

After a very great deal of manipulation (see Appendix B of Islam (1979) for some of the steps) it can be shown that (6.160) holds identically. Thus the system (6.159a–d) (or (6.155a–d)) is consistent. One can therefore choose any three of these equations as the equations determining the functions σ, Ψ and χ. Further, the function $\Omega(\sigma)$, which is an arbitrary function of σ, appears in these equations with its first, second and third derivatives with respect to σ. Thus the system of three equations (any three of (6.159a–d) or of (6.155a–d)) becomes determinant once $\Omega(\sigma)$ is specified. These three equations could equally well have been expressed in terms of the unknowns Ω, Ψ and χ in which case there would have appeared in these equations an arbitrary function $\sigma(\Omega)$ of Ω. This would be similar to the situation in Newton–Maxwell theory (see (6.37)). It is unlikely that one can find a general solution of the system of three of (6.159a–d), since the corresponding Newton–Maxwell equation (6.37) is unsolved. However, there may exist either solutions with no Newtonian analogue, or it may be possible to find the analogue of the Newtonian solution for $m = q$ found in Section (6.3).

One can find exact solutions for differential rotation and vanishing Lorentz force, that is, when both the left and right hand sides of (6.150a, b) vanish separately. These are considered in Van den Bergh and Wils (1984b) and in Islam, Van den Bergh and Wils (1984). However, no solutions exist which are regular and have positive matter density on the axis of rotation. The existence of solutions for differential rotation and cylindrical symmetry was shown by Islam (1983a) and such solutions were found by Van den Bergh and Wils (1984a) some of which are regular on the axis.

In this chapter we have been mainly concerned with finding local solutions to the problem of rotating charged dust. A considerable amount of work remains to be done both on the local and particularly on the global aspect of this problem. As mentioned earlier, exact solutions for well defined energy-momentum tensors are rare and this problem has already yielded a number of interior solutions. We believe it is worthwhile making a detailed study of these and other aspects of this problem.

Appendix

In this appendix we consider an alternative form of Ernst's equation (3.11) for solutions of the form (3.36), that is, rational function solutions which are quotients of complex polynomials in x, y. This form can enable one to search for rational function solutions other than the TS solutions, with the use of algebraic programmes on the computer (see e.g. MacCallum 1984), and also to verify the TS solutions.

From (3.36) we get

$$\xi_x = \frac{N_x + iN'_x}{D + iD'} - \frac{(N + iN')(D_x + iD'_x)}{(D + iD')^2}, \tag{A.1}$$

$$\begin{aligned}\xi_{xx} = \frac{N_{xx} + iN'_{xx}}{D + iD'} &- \frac{2(N_x + iN'_x)(D_x + iD'_x)}{(D + iD')^2} \\ &- \frac{(N + iN')(D_{xx} + iD'_{xx})}{(D + iD')^2} + \frac{2(N + iN')(D_x + iD'_x)^2}{(D + iD')^3}\end{aligned} \tag{A.2}$$

and similar expressions for ξ_y, ξ_{yy}, where $N_x \equiv \partial N/\partial x$, etc. Substituting in (3.11) and multiplying by $(D + iD')^3$ and cancelling a factor $D^2 + D'^2$ in the denominator, we get

$$\begin{aligned}(N^2 + N'^2 &- D^2 - D'^2)\{(x^2 - 1)[(D + iD')^2(N_{xx} + iN'_{xx}) \\ &- 2(D + iD')(N_x + iN'_x)(D_x + iD'_x) \\ &- 2(D + iD')(N + iN')(D_{xx} + iD'_{xx}) + 2(N + iN')(D_x + iD'_x)^2] \\ &+ 2x(D + iD')[(D + iD')(N_x + iN'_x) - (N + iN')(D_x + iD'_x)] - (x \to y)\} \\ = 2(N - iN')&\{(x^2 - 1)[(D + iD')^2(N_x + iN'_x)^2 \\ &- 2(D + iD')(N + iN')(N_x + iN'_x)(D_x + iD'_x) \\ &+ (N + iN')^2(D_x + iD'_x)^2] - (x \to y)\}. \end{aligned} \tag{A.3}$$

In (A.3) the expressions $(x \to y)$ are obtained by substituting y for x in the expressions immediately preceding the signs $(x \to y)$. Resolving (A.3) into real and imaginary parts, we get the following two equations:

$$(N^2 + N'^2 - D^2 - D'^2)\{(x^2 - 1)[(D^2 - D'^2)N_{xx} - 4DD'N'_{xx}$$
$$- 2D(N_xD_x - N'_xD'_x) + 2D'(N_xD'_x + N'_xD_x) - D(ND_{xx} - N'D'_{xx})$$
$$+ D'(ND'_{xx} + N'D_{xx}) + 2N(D_x^2 - D_x'^2) - 4N'D_xD'_x]$$
$$+ 2x[(D^2 - D'^2)N_x - 4DD'N'_x - D(ND_x - N'D'_x)$$
$$+ D'(ND'_x + N'D_x)] - (x \to y)\}$$
$$= [2N(D^2 - D'^2)(N_x^2 - N_x'^2) - 8NDD'N_xN'_x$$
$$+ 4N'(D^2 - D'^2)N_xN'_x + 4N'DD'(N_x^2 - N_x'^2)$$
$$- 4N(DD_x - D'D'_x)(NN_x - N'N'_x) + 4N(DD'_x + D'D_x)(NN'_x + N'N_x)$$
$$- 4N'(DD_x - D'D'_x)(NN'_x + N'N_x) - 4N'(DD'_x + D'D_x)(NN_x - N'N'_x)$$
$$+ 2N(N^2 - N'^2)(D_x^2 - D_x'^2) - 8N^2N'D_xD'_x + 4N'(N^2 - N'^2)D_xD'_x$$
$$+ 4NN'^2(D_x^2 - D_x'^2)](x^2 - 1) - (x \to y), \qquad\qquad (A.4)$$

$$(N^2 + N'^2 - D^2 - D'^2)\{(x^2 - 1)[(D^2 - D'^2)N'_{xx} + 2DD'N_{xx}$$
$$- 2D(N_xD'_x + N'_xD_x) - 2D'(N_xD_x - N'_xD'_x) - D(ND'_{xx} + N'D_{xx})$$
$$- D'(ND_{xx} - N'D'_{xx}) + 2N'(D_x^2 - D_x'^2) + 4ND_xD'_x]$$
$$+ 2x[(D^2 - D'^2)N'_x + 2DD'N_x - D(ND'_x + N'D_x)$$
$$- D'(ND_x - N'D'_x)] - (x \to y)\}$$
$$= [4N(D^2 - D'^2)N_xN'_x + 4NDD'(N_x^2 - N_x'^2)$$
$$- 2N'(D^2 - D'^2)(N_x^2 - N_x'^2) + 8N'DD'N_xN'_x$$
$$- 4N(DD_x - D'D'_x)(NN'_x + N'N_x) - 4N(DD'_x + D'D_x)(NN_x - N'N'_x)$$
$$+ 4N'(DD_x - D'D'_x)(NN_x - N'N'_x) - 4N'(DD'_x + D'D_x)(NN'_x + N'N_x)$$
$$+ 4N(N^2 - N'^2)D_xD'_x + 4N^2N'(D_x^2 - D_x'^2) - 2N'(N^2 - N'^2)$$
$$\cdot(D_x^2 - D_x'^2) + 8NN'^2D_xD'_x](x^2 - 1) - (x \to y) \qquad (A.5)$$

where $(x \to y)$ is defined as before, noting that, as before, x has to be replaced by y also in the suffixes.

References

Banerjee, A. & Banerji, S., 1968, *J. Phys. A.*, **1**, 188.

Banerjee, A., Chakravarty, N. & Dutta Choudhury, S.B., 1976, *Aust. J. Phys.*, **29**, 119.

Bardeen, J.M., 1971, *Astrophys. J.*, **167**, 425.

Bardeen, J.M. & Wagoner, R.V., 1969, *Astrophys. J.*, **158**, L65.

Belinsky, V.A. & Zakharov, V.E., 1978, *Sov. Phys., JETP*, **48**, 985.

Belinsky, V.A. & Zakharov, V.E., 1979, *Sov. Phys., JETP*, **50**, 1.

Boachie, L.A. & Islam, J.N., 1983, *Phys. Lett.*, **A93**, 321.

Bonnor, W.B., 1966, *Z. Phys.*, **190**, 444.

Bonnor, W.B., 1973, *Commun. Math. Phys.*, **34**, 77.

Bonnor, W.B., 1980a, *J. Phys. A: Math Gen.*, **13**, 2121.

Bonnor, W.B., 1980b, *J. Phys. A: Math. Gen.*, **13**, 3465.

Bose, S.K., 1980, *An Introduction to General Relativity*, Wiley Eastern, New Delhi.

Boyer, R.H., & Lindquist, R.W., 1967, *J. Math. Phys.*, **8**, 265.

Carter, B., 1968, *Phys. Rev.*, **174**, 1559.

Carter, B., 1969, *J. Math. Phys.*, **10**, 70.

Carter, B., 1970, *Commun. Math. Phys.*, **17**, 223.

Carter, B., 1971, *Phys. Rev. Lett.*, **26**, 331.

Carter, B., 1972, in *Black Holes* (Les Houches Lectures, 1972), Gordon and Breach, New York.

Chakraborty, S.K., 1980, *Gen. Rel. Grav.*, **12**, 925.

Chandrasekhar, S., 1969, *Ellipsoidal Figures of Equilibrium*, Yale University Press, New Haven.

Chandrasekhar, S., 1971, *Astrophys. J.*, **167**, 447.

Chandrasekhar, S., 1978, *Proc. Roy. Soc. Lond.*, **A358**, 405.

Chandrasekhar, S., 1983, *The Mathematical Theory of Black Holes*, Oxford University Press.

Cosgrove, C.M., 1977, *J. Phys. A: Math. Gen.*, **10**, 1481.

Cosgrove, C.M., 1978, *J. Phys. A: Math. Gen.*, **11**, 2389, 2405.

Cosgrove, C.M., 1980, *J. Math. Phys.*, **21**, 2417.

Cosgrove, C.M., 1982, *J. Math. Phys.*, **23**, 615.

Curzon, H.E.J., 1924, *Proc. Lond. Math. Soc.*, **23**, 477.

De, U.K. & Raychaudhuri, A.K., 1968, *Proc. Roy. Soc. Lond.*, **A303**, 97.

Economou, J.E., 1976, *J. Math. Phys.*, **17**, 1095.
Ernst, F.J., 1968a, *Phys. Rev.*, **167**, 1175.
Ernst, F.J., 1968b, *Phys. Rev.*, **168**, 1415.
Ernst, F.J., 1976, *J. Math. Phys.*, **17**, 1091.
Frehland, E., 1971, *Commun. Math. Phys.*, **23**, 127.
Geroch, R., 1972, *J. Math. Phys.*, **13**, 394.
Gibbons, G.W. & Russell-Clark, R.A., 1973, *Phys. Rev. Lett.*, **30**, 398.
Gödel, K., 1949, *Rev. Mod. Phys.*, **21**, 447.
Harrison, B.K., 1978, *Phys. Rev. Lett.*, **41**, 1197.
Hartle, J.B. & Hawking, S.W., 1972, *Commun. Math. Phys.*, **26**, 87.
Hauser, I. & Ernst, F.J., 1979, *Phys. Rev.*, **D20**, 1783.
Hauser, I. & Ernst, F.J., 1980, *J. Math. Phys.*, **21**, 1126.
Hawking, S.W., 1972, *Commun. Math. Phys.*, **25**, 152.
Hawking, S.W. & Ellis, G.F.R., 1973, *The Large Scale Structure of Spacetime*, Cambridge University Press.
Herlt, E., 1978, *Gen. Rel. Grav.*, **9**, 711.
Hernandez, W.C., 1967, *Phys. Rev.*, **159**, 1070.
Hoenselaers, C., 1979, *J. Math. Phys.*, **20**, 2526.
Hoenselaers, C. & Ernst, F.J., 1983, *J. Math. Phys.*, **24**, 1817.
Hoenselaers, C. Kinnersley, W. & Xanthopoulos, B.C., 1979, *J. Math. Phys.*, **20**, 2530.
Hoenselaers, C. & Vishveshwara, C.V., 1979a, *Gen. Rel. Grav.*, **10**, 43.
Hoenselaers, C. & Vishveshwara, C.V., 1979b, *J. Phys. A: Math. Gen.*, **12**, 209.
Hori, S., 1978, *Progr. Theor. Phys.*, **59**, 1870.
Islam, J.N., 1976a, *Math. Proc. Camb. Phil. Soc.*, **79**, 161.
Islam, J.N., 1976b, *Gen. Rel. Grav.*, **7**, 669.
Islam, J.N., 1976c, *Gen, Rel. Grav.*, **7**, 809.
Islam, J.N., 1977, *Proc. Roy. Soc. Lond.*, **A353**, 532.
Islam, J.N., 1978a, *Proc. Roy. Soc. Lond.*, **A362**, 329.
Islam, J.N., 1978b, *Gen. Rel. Grav.*, **9**, 687.
Islam, J.N., 1979, *Proc. Roy. Soc. Lond.*, **A367**, 271.
Islam, J.N., 1980, *Proc. Roy. Soc. Lond.*, **A372**, 111.
Islam, J.N., 1983a, *Proc. Roy. Soc. Lond.*, **A385**, 189.
Islam, J.N., 1983b, *Proc. Roy. Soc. Lond.*, **A389**, 291.
Islam, J.N., 1983c, *Phys. Lett.*, **94A**, 421.
Islam, J.N., 1984, in *Classical General Relativity*, eds. W.B. Bonnor, J.N. Islam & M.A.H. MacCallum, Cambridge University Press.
Islam, J.N., Van den Bergh, N. & Wils, P., 1984, in press.
Israel, W., 1967, *Phys. Rev.*, **164**, 1776.
Israel, W. & Spanos, J.T.J., 1973, *Lett. Nuovo Cim.*, **7**, 245.
Israel, W. & Wilson, G.A., 1972, *J. Math. Phys.*, **13**, 865.
Jordan, R., 1983, unpublished.
Kerr, R.P., 1963, *Phys. Rev. Lett.*, **11**, 237.
Kinnersley, W., 1977, *J. Math. Phys.*, **18**, 1529.
Kinnersley, W. & Chitre, D.M., 1977, *J. Math. Phys.*, **18**, 1538.
Kinnersley, W. & Chitre, D.M., 1978a, *J. Math. Phys.*, **19**, 1037.
Kinnersley, W. & Chitre, D.M., 1978b, *J. Math. Phys.*, **19**, 1926.
Kinnersley, W. & Kelley, E.F., 1974, *J. Math. Phys.*, **15**, 2121.

Kloster, S. & Das, A., 1977, *J. Math. Phys.*, **18**, 2191.
Kramer, D., Stephani, H., Herlt, E. & MacCallum, M.A.H., 1980, *Exact Solutions of Einstein's Field Equations*, Cambridge University Press.
Kruskal, M.D., 1960, *Phys. Rev.*, **119**, 1743.
Landau, L.D. & Lifshitz, E.M., 1975, *The Classical Theory of Fields*, 4th eds, Pergamon Press, Oxford.
Lewis, T., 1932, *Proc. Roy. Soc. Lond.*, **A136**, 176.
Lynden-Bell, D. & Pineault, S., 1978, *Mon. Not. Roy. Astron. Soc.*, **185**, 679, 695.
MacCallum, M.A.H., 1984, in *Classical General Relativity*. eds. W.B. Bonner, J.N. Islam & M.A.H. MacCallum, Cambridge University Press.
Majumdar, S.D., 1947, *Phys. Rev.*, **72**, 390.
Misner, C.M., Thorne, K.S. & Wheeler, J.A., 1973, *Gravitation*, W.H. Freeman, San Francisco.
Newman, E.T., Couch, E., Chinnapared, K. Exton, A., Prakash, A. & Torrence, R., 1965, *J. Math. Phys.*, **6**, 918.
Neugebauer, G., 1979, *J. Phys. A: Math. Gen.*, **12**, L67.
Papapetrou, A., 1947, *Proc. Roy. Irish Acad.*, **A51**, 191.
Papapetrou, A., 1953, *Ann. Physik*, **12**, 309.
Papapetrou, A., 1966, *Ann. Inst. H. Pincaré*, **A4**, 83.
Perjés, Z., 1971, *Phys. Rev. Lett.*, **27**, 1668.
Raychaudhuri, A.K., 1982, *J. Phys. A: Math. Gen.*, **15**, 831.
Raychaudhuri, A.K. & De, U.K., 1970, *J. Phys. A: Gen. Phys.*, **3**, 263.
Robinson, D.C., 1975, *Phys. Rev. Lett.*, **34**, 905.
Som, M.M. & Raychaudhuri, A.K., 1968, *Proc. Roy. Soc. Lond.*, **A304**, 8.
Som, M.M., Teixeira, A.F.F. & Wolk, I., 1976, *Gen. Rel. Grav.*, **7**, 263.
Tanabe, Y., 1979, *J. Math. Phys.*, **20**, 1486.
Thirring, H. & Lense, J., 1918, *Phys. Z.*, **19**, 156.
Thorne, K.S., 1971, in *Gravitation and Cosmology*, ed. R.K. Sachs, Academic Press, New York.
Tipler, F.J., 1974, *Phys. Rev.*, **D9**, 2203.
Tomimatsu, A. & Sato, H., 1972, *Phys. Rev. Lett.*, **29**, 1344.
Tomimatsu, A. & Sato, H., 1973, *Progr. Theor. Phys.*, **50**, 95.
Van den Bergh, N. & Wils, P., 1983, *J. Phys. A: Math. Gen.*, **16**, 3843.
Van den Bergh, N. & Wils, P., 1984*a*, *b*, *Classical and Quantum Gravity*, **1**, 199, and in press.
Van den Bergh, N. & Wils, P., 1984*c*, in *Classical General Relativity*, eds. W.B. Bonnor, J.N. Islam & M.A.H. MacCallum, Cambridge University Press.
Van den Bergh, N. & Wils, P., 1984*d*, University of Antwerp preprint.
Van Stockum, W., 1937, *Proc. Roy. Soc. Edinb.*, **57**, 135.
Vishveshwara, C.V. & Winicour, J., 1977, *J. Math. Phys.*, **18**, 1280.
Walker, A.G., 1935, *Proc. Edin. Math. Soc.*, **iv**, 107.
Weinberg, S., 1972, *Gravitation and Cosmology*, Wiley, New York.
Weyl, H., 1917, *Ann. Physik*, **54**, 117.
Winicour, J., 1975, *J. Math. Phys.*, **16**, 1806.
Yamazaki, M., 1977*a*, *Progr. Theor. Phys.*, **57**, 1951.
Yamazaki, M., 1977*b*, *J. Math. Phys.*, **18**, 2502.

Index